MECHANICAL TECHNOLOGY IN AGRICULTURE

MECHANICAL TECHNOLOGY IN AGRICULTURE

Dr. D.V. Bhagat

RANDOM PUBLICATIONS

NEW DELHI - 110 002 (INDIA)

Mechanical Technology in Agriculture

ISBN 978-93-51116-80-6

Published in 2015 in India by

RANDOM PUBLICATIONS

4376-A/4B, Gali Murari Lal, Ansari Road
New Delhi-110 002
Phone: +9111-43580356, 23289044
E-mail: randomexports@gmail.com; sales@randompublications.com; info@randompublications.com

Reprint 2021

Type Setting by: Friends Media, Delhi-110089
Digitally Printed at : Replika Press Pvt. Ltd.

Preface

Mechanization faces many obstacles before wide adoption is possible in tropical regions. Difficult soils, stones, stumps, abundant labour, resistance from farmers, lack of incentives, lack of skills, lack of capital, low wages, high cost of machines, lack of dealer service, fragmented land ownership, all contribute to slow development of mechanization. Tropical soils differ markedly from those in the countries that manufacture land-preparation machinery, making adaptation of new design necessary. The encountering of stones, wood, trash, and termite mounds causes machines to break down. Depressing climatic conditions reduce the performance of the machine operators. Tropical farm regions are notoriously irregular or mountainous, impeding intensive machine culture. Historically, agriculture meant the practice of farming or growing animals and plants. Today, it is better defined as the science and art of growing animals and plants while protecting the environment, accompanied by related activities involving supplies, services, mechanics, products, processing, and marketing. Actually, modern agriculture covers so many activities that a simple definition is not adequate. Agricultural mechanics has been a major component in the modernization of agriculture. Agricultural mechanization is the power and machinery used to produce food and fibre from the land. Early equipment was made of wood and powered by hand or animals. One of the key early innovations was Ely Whitney's gin, a machine to remove seeds from cotton. The limits of human and animal power prevented most production units from growing very large.

Many early machines were designed and made by local blacksmiths. The plow and the grain harvester are examples of such early machines. They represent the new technology and advancement of their time. These two machines, created in the 1830s, revolutionized mechanical technology in agriculture. Farmers and farm machinery continued to evolve. The threshing machine replaced hand threshing, which would be replaced by the combine. Mechanization has played an important role in the productivity of agriculture. American inventors have created many of the world's important agricultural machines. At the end of the twentieth century, information technology and precision techniques were increasingly accepted. Farmers began using satellite technology to track and plan their agricultural practices. The use of conservation tillage methods to combat erosion continued to rise. The first biotech crops—soybeans and cotton—appeared, reducing the dependence on pesticides. Environmentally controlled livestock production facilities became acceptable. The greenhouse is typically a structure whose roof and sides are transparent or translucent, permitting a sufficient quality and quantity of solar radiation to enter the structure for photosynthesis. It allows the growing of crops independently of the outside climate, since its interior temperature and humidity can be controlled.

This book presents the practical applications and theoretical understandings necessary for both occupational success and advanced study in agricultural technology for students and people engaged in the field.

I thank all members of my team who have helped in the preparation of the book. My special thanks go to "Random Publications" who have published the book.

— Dr. D.V. Bhagat

Contents

Chapter 1

Introduction

Agricultural Technology

Agricultural technology, application of techniques to control the growth and harvesting of animal and vegetable products.

Soil Preparation

Mechanical processing of soil so that it is in the proper physical condition for planting is usually referred to as tilling; adding nutrients and trace elements is called fertilizing. Both processes are important in agricultural operations.

Tilling

Tillage is the manipulation of the soil into a desired condition by mechanical means; tools are employed to achieve some desired effect (such as pulverization, cutting, or movement). Soil is tilled to change its structure, to kill weeds, and to manage crop residues. Soil-structure modification is often necessary to f]acilitate the intake, storage, and transmission of water and to provide a good environment for seeds and roots. Elimination of weeds is important, because they compete for water, nutrients, and light. Crop residues on the surface must be managed in order to provide conditions suitable for seeding and cultivating a crop.

Generally speaking, if the size of the soil aggregates or particles is satisfactory, preparation of the seedbed will consist only of removing weeds and the management of residues. Unfortunately, the practices associated with planting, cultivating, and harvesting usually cause destruction of soil structure. This leaves preparation of the seedbed as the best opportunity to create desirable structure, in which large and stable pores extend from the soil surface to the water table or drains, ensuring rapid infiltration and drainage of excess or free water and promoting aeration of the subsoil. When these large pores are interspersed with small ones, the soil will retain and store moisture also.

Seedbed-preparation procedures depend on soil texture and the desired change in size of aggregates. In soils of coarse texture, tillage will increase aggregate size, provided it is done when only the small pores are just filled with water; tillage at other than this ideal moisture will make for smaller aggregates. By contrast, fine-textured soils form clods; these require breakage into smaller units by weathering or by machines. If too wet or too dry, the power requirements for shattering dry clods or cutting wet ones are prohibitive when using tillage alone. Thus, the farmer usually attempts tillage of such soils only after a slow rain has moistened the clods and made them friable.

Some soils require deepening of the root zone to permit increased rate of water intake and improved storage. Unfavourable aeration in zones of poor drainage also limits root development and inhibits use of water in the subsoil.

Tillage, particularly conventional plowing, may create a hardpan, or plow sole; that is, a compacted layer just below the zone disturbed by tillage. Such layers are more prevalent with increasing levels of mechanization; they reduce crop yields and must be shattered, allowing water to be stored in and below the shattered zone for later crops.

Primary Tillage Equipment

Equipment used to break and loosen soil for a depth of six to 36 inches (15 to 90 centimetres) may be called primary tillage equipment. It includes moldboard, disk, rotary, chisel, and subsoil plows.

The moldboard plow is adapted to the breaking of many soil types. It is well suited for turning under and covering crop residues. There are hundreds of different designs, each intended to function best in performing certain tasks in specified soils. The part that breaks the soil is called the bottom or base; it is composed of the share, the landside, and the moldboard.

When a bottom turns the soil, it cuts a trench, or furrow, throwing to one side a ribbon of soil that is called the furrow slice. When plowing is started in the middle of a strip of land, a furrow is plowed across the field; on the return trip, a furrow slice is lapped over the first slice. This leaves a slightly higher ridge than the second, third, and other slices. The ridge is called a back furrow. When two strips of land are finished, the last furrows cut leave a trench about twice the width of one bottom, called a dead furrow. When land is broken by continuous lapping of furrows, it is called flat broken. If land is broken in alternate back furrows and dead furrows, it is said to be bedded or listed.

Different soils require different-shaped moldboards in order to give the same degree of pulverization of the soil. Thus, moldboards are divided into several different classes, including stubble, general-purpose, general-purpose for clay and stiff-sod soil, slat, blackland, and chilled general-purpose. The blackland bottom is used, for example, in areas in which the soil does not scour easily; that is, where the soil does not leave the surface of the emerging plow clean and polished.

The share is the cutting edge of the moldboard plow. Its configuration is related to soil type, particularly in the down suction, or concavity, of its lower surface. Generally, three degrees of down suction are recognized: regular for light soil, deep for ordinary dry soil, and double-deep for clay and gravelly soils. In addition, the share has horizontal suction, which is the amount its point is bent out of line with the landside. Down suction causes the plow to penetrate to proper depth when pulled forward, while horizontal suction causes the plow to create the desired width of furrow.

Moldboard-plow bottom sizes refer to width between share wing and the landside. Tractor-plow sizes generally range from 10 to 18 inches (25 to 45 centimetres), although larger, special-purpose types exist.

On modern mechanized farms, plow bottoms are connected to tractors either as trailing implements or integrally. One or more bottoms may be so attached. They are found paired right and left occasionally (two-way), with the advantage of throwing the furrow slice in a constant direction as the turns are made. A variation is the middlebreaker, or lister, which is a bottom equipped with both right- and left-handed moldboards.

The disk plow employs round, concave disks of hardened steel, sharpened and sometimes serrated on the edge, with diameters ranging from 20 to 38 inches (50 to 95 centimetres). It reduces friction by making a rolling bottom in place of a sliding one. Its draft is about the same as that of the moldboard plow. The disk plow works to advantage in situations where the moldboard will not, as in sticky nonscouring soils; in fields with a plow sole; in dry, hard ground; in peat soils; and for deep plowing. The disk-plow bottom is usually equipped with a scraper that aids in pulverizing the furrow slice. Disk plows are either trailed or mounted integrally on a tractor.

The rotary plow's essential feature is a set of knives or tines rotated on a shaft by a power source. The knives chop the soil up and throw it against a hood that covers the knife set. These machines can create good seedbeds, but their high cost and extra power requirement have limited general adoption, except for the small garden tractor.

The chisel plow is equipped with narrow, double-ended shovels, or chisel points, mounted on long shanks. These points rip through the soil and stir it but do not invert and pulverize as well as the moldboard and disk plows. The chisel plow is often used to loosen hard, dry soils prior to using regular plows; it is also useful for shattering plow sole.

Subsoil plows are similar in principle but are much larger, since they are used to penetrate soil to depths of 20 to 36 inches (50 to 90 centimetres). Tractors of 60 to 85 horsepower are required to pull a single subsoil point through a hard soil at a depth of 36 inches. These plows are sometimes equipped with a torpedo-shaped attachment for making subsurface drainage channels.

Secondary Tillage

Secondary tillage, to improve the seedbed by increased soil pulverization, to conserve moisture through destruction of weeds, and to cut up crop residues, is accomplished by use of various types of harrows, rollers, or pulverizers, and tools for mulching and fallowing. Used for stirring the soil at comparatively shallow depths, secondary-tillage equipment is generally employed after the deeper primary-tillage operations; some primary tillage tools, however, are usable for secondary tillage. There are five principal types of harrows: the disk, the spike-tooth, the spring-tooth, the rotary cross-harrow, and the soil surgeon. Rollers, or pulverizers, with V-shaped wheels make a firm and continuous seedbed while crushing clods. These tools often are combined with each other.

When moisture is scarce and control of wind and water erosion necessary, tillage is sometimes carried out in such a way that crop residues are left on the surface. This system is called trash farming, stubble mulch, or subsurface tillage. Principal equipment for subsurface tillage consists of sweeps and rod weeders. Sweeps are V-shaped knives drawn below the surface with cutting planes horizontal. A mounted set of sweeps provided with power lift and depth regulation is often called a field cultivator.

The typical rod weeder consists of a frame with several plowlike beams, each having a bearing at its point. Rods are extended through the bearings, which revolve slowly under power from a drive wheel. The revolving rod runs a few inches below the surface and pulls up vegetative growth; clearance of the growth from the rod is assisted by its rotation. Rod weeders are sometimes attached to chisel plows. Some control of weeds is obtained by tillage that leaves the middles between crop rows loose and cloddy. When a good seedbed is prepared only in the row, the seeded crop can become established ahead of the weeds. Plowing with the moldboard plow buries the weed seeds, retards their sprouting, and tends to reduce the operations needed to control them. If weed infestations become bad, they can be reduced somewhat by undercutting.

Since rainfall amount and distribution seldom match crop needs, farmers usually prefer tillage methods that encourage soil-moisture storage at times when crops are not growing. From the soil-moisture standpoint, any tillage practice that does not control weeds and result in greater moisture intake and retention during the storage period is probably unnecessary or undesirable.

Minimum Tillage

The use of cropping systems with minimal tillage is usually desirable, because intensive tillage tends to break down soil structure. Techniques such as mulching also help prevent raindrops from injuring the surface structure. Excessive tillage leaves the soil susceptible to crusting, impedes water intake, increases runoff, and thus reduces water storage for crop use. Intensive vegetable production in warm climates where three crops per year may be grown on the same land may reduce the soil to a single-grain structure that facilitates surface cementation and poor aeration.

The loosening and granulating actions of plowing may improve soil structure if the plowing is done when the moisture content is optimum; if not so timed, however, plowing can create unfavourable structure. The lifting and inversion of the furrow slice likewise may not always be desirable, because in many cases it is better to leave a trashy surface.

The concept of minimum tillage has received much attention. One type of minimum tillage consists in seeding small grain in sod that has been relatively undisturbed. Narrow slits are cut in the sod and seed and fertilizer placed in the breaks thus formed. Soil normally subject to erosion can be planted to grain this way while still retaining the erosion resistance of the sod. The technique has been successful in preparing winter grazing in southeastern portions of the United States. In another type of minimum tillage, the land is broken and planted without further tillage in seedbed preparation. One approach involves breaking the land and planting seeds in the tractor tracks (wheel-track planting); the tractor weight crushes clods and leaves the seed surrounded by firm soil. Another method consists of mounting a planter behind the plow, thus planting without further traffic and leaving a loose seedbed that is satisfactory in areas where postplanting rains may be heavy. In some areas, where winter rain often comes after wheat is drilled, a rotation of wheat following peas has been successful. After the peas have been harvested, the field is rough plowed, and fall wheat is then drilled in directly. All these methods minimize expense and land preparation, tending to leave the soil rough, which reduces erosion and increases water intake. Somewhat similar systems are employed with row crops, where chemical weed control assists in reducing need for cultivation.

Mulch Tillage

Mulch tillage has been mentioned already; in this system, crop residues are left on the surface, and subsurface tillage leaves them relatively undisturbed. In dryland areas, a maximum amount of mulch is left on the surface; in more humid regions, however, some of the mulch is buried. Planting is accomplished with disk openers that go through several inches of mulch. Since mulch decomposition may deprive the crop of nitrogen, extra fertilizer is often placed below the mulch in humid areas. In rainy sections, intercropping extends the protection against erosion provided by mulches. Intercrops are typically small grains or sod crops such as alfalfa or clover grown between the rows of a field crop that reach maturity shortly after the field crop has been established and furnish mulch cover for a long time.

If growth of the intercrop competes with the main crop for moisture and nutrients, that growth may be killed at seeding time or soon thereafter by undercutting with sweeps.

Tillage in dry areas must make maximum use of scanty rainfall. The lister (double-mold board) plow, or middlebreaker, is here used to make water-impounding ridges that promote infiltration. The special problems of dryland farming will be considered below.

Fertilizing and Conditioning the Soil

Soil fertility is the quality of a soil that enables it to provide compounds in adequate amounts and proper balance to promote growth of plants when other factors (such as light, moisture, temperature, and soil structure) are favourable. Where fertility of a soil is not good, natural or manufactured materials may be added to supply the needed plant nutrients; these are called fertilizers, although the term is generally applied to largely inorganic materials other than lime or gypsum. Fertilizer grade is a conventional expression that indicates the percentage of plant nutrients in a fertilizer; thus, a 10–20–10 grade contains 10 percent nitrogen, 20 percent phosphoric oxide, and 10 percent potash. The green plant, however, requires more nutrients than these.

Essential Plant Nutrients

In total, the plant has need of at least 16 elements, of which the most important are carbon, hydrogen, oxygen, nitrogen, phosphorus, sulfur, potassium, calcium, and magnesium.

The plant obtains carbon and hydrogen dioxide from the atmosphere; other nutrients are taken up from the soil. Although the plant contains sodium, iodine, and cobalt, these are apparently not essential. This is also true of silicon and aluminum.

Overall chemical analyses indicate that the total supply of nutrients in soils is usually high in comparison with the requirements of crop plants. Much of this potential supply, however, is bound tightly in forms that are not released to crops fast enough to give satisfactory growth. Because of this, the farmer is interested in measuring the available nutrient supply as contrasted to the total quantities. This point will be considered later.

The solid content of soils is broadly classified as organic and inorganic. Materials of organic origin range from fresh plant tissue to the more or less stable black or brown degradation product (humus) formed by biological decay. The organic matter is a potential source of nitrogen, phosphorus, and sulfur; it contains more than 95 percent of the total nitrogen, 5 to 60 percent of the total phosphorus, and 10 to 80 percent of the total sulfur.

These three elements are cycled through the entire environment of living things (the biosphere). The soil organic matter can be considered as one of the storage points in these cycles. Where nonlegumes are grown in the absence of fertilizer or manures, the crop must gain its nitrogen supply from the organic matter; only a part, however, of the needed phosphorus and sulfur is so supplied.

The inorganic or mineral fraction, which comprises the bulk of most soils, is derived from rocks and their degradation products. The power to supply plant nutrients is much greater in the larger particles, sand and silt, than in the fine particles, or clay. The minerals that comprise the sand and silt in soil contain most of the elements essential for plant growth as a part of their structure. The difficulty is that these minerals decompose so slowly in soil that the rate of supply of the nutrient elements is usually insufficient for good growth of plants.

When the available supply of a given nutrient becomes depleted, its absence becomes a limiting factor in plant growth, and the addition of this nutrient to the soil will increase yields of dry matter. Excessive quantities of some nutrients may cause decrease in yield, however.

Determining Nutrient Needs

Determination of a crop's nutrient needs is an essential aspect of fertilizer technology. The appearance of a growing crop may indicate need of fertilizer; in some plants, however, the need for more or different nutrients may not be easily observable. If such a problem exists, its nature must be diagnosed, the degree of deficiency must be determined, and the amount and kind of fertilizer needed for a given yield must be found. There is no substitute for detailed examination of plants and soil conditions in the field, followed by simple fertilizer tests, quick tests of plant tissues, and analysis of soils and plants.

Sometimes plants show symptoms of poor nutrition. Chlorosis (general yellow or pale-green colour), for example, indicates lack of sulfur and nitrogen. Iron deficiency produces white or pale-yellow tissue. Symptoms can be misinterpreted, however. Plant disease can produce appearances resembling mineral deficiency, as can various organisms. Drought or improper cultivation or fertilizer application each may create deficiency symptoms.

After field diagnosis, the conclusions may be confirmed by experiments in a greenhouse or by making strip tests in the field. In strip tests, the fertilizer elements suspected of being deficient are added, singly or in combination, and the resulting plant growth observed. Next, it is necessary to determine the extent of the deficiency.

An experiment in the field can be conducted by adding nutrients to the crop at various rates. The resulting response of yield in relation to amount of nutrient supplied will indicate the supplying power of the unfertilized soil in terms of bushels or tons of produce. If the increase in yield is large, this practice will show that the soil has too little of a given nutrient. Such field experiments may not be practical, because they can cost too much in time and money. Soil-testing laboratories are available in most areas; they conduct chemical soil tests to estimate the availability of nutrients. Commercial soil-testing kits give results that may be very inaccurate, depending on techniques and interpretation. Actually, the most accurate system consists of laboratory analysis of the nutrient content of plant parts, such as the leaf. The results, when correlated with yield response to fertilizer application in field experiments, can give the best estimate of deficiency. Further development of remote sensing techniques, such as infrared photography, are under study and may ultimately become the most valuable technique for such estimates.

The Economics of Fertilizers

The practical goal is to determine how much nutrient material to add. Since the farmer wants to know how much profit to expect if he buys extra fertilizer, the tests are interpreted as an estimation of increased crop production that will result from nutrient additions. Cost of nutrients must be balanced against value of crop or even against alternative procedures, such as investing the money in something else with a greater potential return. The law of diminishing returns is well exemplified in fertilizer technology. Past a certain point, equal inputs of chemicals produce less and less yield increase. The goal of the farmer is to use fertilizer in such a way that the most profitable rate is employed.

Fertilizers can aid in making profitable changes in farming. Operators can reduce costs per unit of production and increase the margin of return over total cost by increasing rates of application of fertilizer on principal cash and feed crops. They are then in a position to invest in soil conservation and other improvements that are needed when shifting acreage from surplus crops to other uses.

Farm Manure

Among sources of organic matter and plant nutrients, farm manure has been of major importance in past years. Manure is understood to mean the refuse from stables and barnyards, including both excreta and straw or other bedding material, while the term fertilizer refers to chemicals. Large amounts of manure are produced by livestock; such manure has value in maintaining and improving soil because of the plant nutrients, humus, and organic substances contained in it.

As manure must be managed carefully in order to derive the most benefit from it, some farmers may be unwilling to expend the necessary time and effort. Manure must be carefully stored to minimize loss of nutrients, particularly nitrogen. It must be applied to the right kind of crop at the proper time. Also, additional fertilizer may be needed, such as phosphoric oxide, in order to gain full value of the nitrogen and potash that are contained in manure.

Manure is fertilizer graded as approximately 0.5–0.25–0.5 (percentages of nitrogen, phosphoric oxide, and potash), with at least two-thirds of the nitrogen in slow-acting forms. Commercial fertilizer equivalent to one ton (900 kilograms) of average manure can be purchased at a fairly low price. Furthermore, the expense of applying

100 pounds (45 kilograms) of 10–5–10 fertilizer is much less than the cost of applying 20 times as much manure. On properly tilled soils, the returns from fertilizer usually will be greater than from an equivalent amount of manure. The application of manure to a crop cannot be controlled as readily as can granulated fertilizer. In general, manure does not provide all the plant nutrients needed and fails to provide any that cannot be supplied by artificial fertilizers. Thus, there is a tendency to discount the value of manure as fertilizer. In underdeveloped countries, however, where artificial fertilizer may be costly or unavailable and where labour is relatively cheap, manure is attractive as a fertilizer.

The main benefits of manure are indirect. It supplies humus, which improves the soil's physical character by increasing its capacity to absorb and store water, by enhancement of aeration, and by favouring the activities of lower organisms. Manure incorporated into the topsoil will help prevent erosion from heavy rain and slow down evaporation of water from the surface. In effect, the value of manure as a mulching material may be greater than is its value as a source of essential plant nutrients.

Green Manuring

In reasonably humid areas, the practice of green manuring can improve yield and soil qualities. A green-manure crop is grown and plowed under for its beneficial effects, although during its growth it may be grazed. These green crops are usually annuals, either grasses or legumes, whose roots bear nodule bacteria capable of fixing atmospheric nitrogen. Among the advantages of green-manure crops are the addition of nitrogen to the soil, increase in general fertility level, reduction of erosion, improvement of physical condition, and reduction of nutrient loss from leaching. Disadvantages include the chance of not obtaining a satisfactory growth; the possibility that the cost of growing the manure crop may exceed the cost of applying commercial nitrogen; possible increases in disease, insects, and nematodes (parasitic worms); and possible exhaustion of soil moisture by the crop.

Green-manure crops are usually planted in the fall and turned under in the spring before the summer crop is sown. Their value as a source of nitrogen, particularly that of the legumes, is unquestioned for certain crops such as potatoes, cotton, and corn (maize); for other crops, such as peanuts (groundnuts; themselves legumes), the practice is questionable. Farmers are gradually turning away from growing green-manure crops except where the crop may also serve as winter cover for the land.

Compost, Peat, and Sludge

Compost, peat, and sludge are used in agriculture and gardening as soil amendments rather than as fertilizers, because they have a low content of plant nutrients. They may be incorporated into the soil or mulched on the surface. Heavy rates of application are common.

Compost, or synthetic manure, is basically a mass of rotted organic matter made from waste-plant residues. Addition of nitrogen during decomposition is usually advisable. The result is a crumbly material that when added to soil does not compete with the crop for nitrogen. When properly prepared, it is free of obnoxious odours. Composts commonly contain about 2 percent nitrogen, 0.5 to 1 percent phosphorus, and about 2 percent potassium; if phosphate and potash are added while composting, those values are higher. The nitrogen of compost becomes available slowly and never approaches that available from inorganic sources. This slow release of nitrogen reduces leaching and extends availability over the whole growing season. Composts are essentially fertilizers with low nutrient content, which explains why large amounts are applied. The maximum benefits of composts on soil structure (better aggregation, pore spacing, and water storage) and on crop yield usually occur after several years of use.

In practical farming, the use of composted plant residues must be compared to the use of fresh residues. More beneficial soil effects usually accrue with less labour by simply turning under fresh residues; also, since one-half the organic matter is lost in composting, fresh residues applied at the same rate will cover twice the area that composted residues would cover. In areas where commercial fertilizers are expensive, labour is cheap, and implements are simple, however, composting meets the needs and is a logical practice.

Peat, composed of prehistoric plant remains that have accumulated under airless conditions in bogs, is a widely used organic soil amendment. Peat moss, the remains of sphagnum plants, is probably its most common form; it contains less than 1 percent nitrogen, with phosphorus and potassium below 0.1 percent. It is highly acid, with pH between 3 and 4.5 (a pH value of 7 is neutral and one above 7 basic). Peat improves the water-storage capability of soils and gives better structure to fine soils. Heavy applications of peat is usually the practice. It is used mostly by specialty-crop producers and on lawns and gardens.

Sewage sludge is the solid material remaining from the treatment of sewage. Its value for soil improvement depends on the method used for treating the sewage. Activated sludge, which results from aerobic (oxygen) treatment, contains 5 to 6 percent nitrogen and 1 to 3.5 percent of phosphorus. After suitable processing, it is sold as fertilizer and soil amendment for use on lawns, parks, and golf courses. It is rarely used in farming.

Liming

Liming to reduce soil acidity is practiced extensively in humid areas where rainfall leaches calcium and magnesium from the soil, thus creating an acid condition. Calcium and magnesium are major plant nutrients supplied by liming materials. Ground limestone is widely used for this purpose; its active agent, calcium carbonate, reacts with the soil to reduce its acidity. The calcium is then available for plant use. The typical limestones, especially dolomitic, contain magnesium carbonate as well, thus also supplying magnesium to the plant.

Another liming material is basic slag, a by-product of steel manufacture; its active ingredient is calcium silicate. Marl and chalk are soft, impure forms of limestone and are sometimes used as liming materials, as are oyster shells. Calcium sulfate (gypsum) and calcium chloride, however, are unsuitable for liming, for, although their calcium is readily soluble, they leave behind a residue that is harmful.

Lime is applied by mixing it uniformly with the surface layer of the soil. It may be applied at any time of the year on land plowed for spring crops or winter grain or on permanent pasture. After application, plowing, disking, or harrowing will mix it with the soil. Such tillage is usually necessary, because calcium migrates slowly downward in most soils. Lime is usually applied by trucks specially equipped and owned by custom operators.

Methods of Application

Fertilizers may be added to soil in solid, liquid, or gaseous forms, the choice depending on many factors. Generally, the farmer tries to obtain satisfactory yield at minimum cost in money and labour.

Manure can be applied as a liquid or a solid. When accumulated as a liquid from livestock areas, it may be stored in tanks until needed and then pumped into a distributing machine or into a sprinkler irrigation system. The method reduces labour, but the noxious odours are objectionable. The solid-manure spreader conveys the material to the field, shreds it, and spreads it uniformly over the land. The process can be carried out during convenient times, including winter, but rarely when the crop is growing.

Application of granulated or pelleted solid fertilizer has been aided by improved equipment design. Such devices, depending on design, can deposit fertilizer at the time of planting, side-dress a growing crop, or broadcast the material. Fertilizer attachments are available for most tractor-mounted planters and cultivators and for grain drills and some types of plows. They deposit fertilizer with the seed when planted, without damage to the seed, yet the nutrient is readily available during early growth. Placement of the fertilizer varies according to the types of crops; some crops require banding above the seed, while others are more successful when the fertilizer band is below the seed.

The use of liquid and ammonia fertilizers is growing, particularly of anhydrous ammonia, which is handled as a liquid under pressure but changes to gas when released to atmospheric pressure. Anhydrous ammonia, however, is highly corrosive, inflammable, and rather dangerous if not handled properly; thus, application equipment is quite specialized. Typically, the applicator is a chisel-shaped blade with a pipe mounted on its rear side to conduct the ammonia five to six inches (13 to 15 centimetres) below the surface. Pipes are fed from a pressure tank mounted above. Mixed liquid fertilizers containing nitrogen, phosphorus, and potassium may be applied directly to the surface—by field sprayers where close-growing crops are raised. Large areas can be covered rapidly by use of aircraft, which can distribute both liquid and dry fertilizer.

The Future for Fertilizers

Future trends in fertilizer technology may be predicted by extrapolating from current developments. Mixtures and materials with high percentages of plant nutrients will dominate the field. Better ways of providing nitrogen, the most expensive of the three major nutrients, will be forthcoming, including increased use of anhydrous ammonia, ammonium nitrate, and urea. Nonleachable nitrogen, for example, can be obtained through the urea–formaldehyde (ureaform) reaction, and ammonium metaphosphate offers a concentrated liquid product. Micronutrients, or trace elements, specific to particular geographical areas will come into increasing use, as will custom mixing and bulk selling of mixtures containing several nutrients based on reliable soil and plant data.

"Complete environment" seeding in which seed, fertilizer, and water are incorporated in a biodegradable (decomposable in the soil) tape may come into use; with the tape planted, no further fertilizer or water will be needed until growth is well established. Such techniques using biodegradable tapes have already been developed on a small scale for use by home gardeners. Finally, larger and more precise fertilizing machines will be developed and adopted.

Factors in Cropping

Cropping Systems

The kind and sequence of crops grown over a period of time on a given area of soil can be described as the cropping system. It may be a pattern of regular rotation of different crops or one of growing only one crop year after year on the same area.

Crop Rotation

Early agricultural experiments showed the value of crop rotations that included a legume sod crop in the regular sequence. Such a system generally maintains productivity, aids in keeping soil structure favourable, and tends to reduce erosion. Alfalfa, sweet clover, red clover, and Ladino clover are considered effective for building up nitrogen. Some legumes, however, do not leave nitrogen behind in the soil because it is deposited as protein in the harvested seed; soybeans are an example. Turning under the top growth of a legume aids in adding nitrogen. Though yields of grains are higher when they are rotated with legumes, it is difficult to determine how much of the improvement depends on the nitrogen added by the legume and how much on improved soil structure or fewer insects and disease.

The determination of the best rotation depends upon whether the crops compete with each other (*i.e.,* if growing one crop lowers the yield of its successor) or complement each other; and the output of one crop on a given acreage leads to increased output of the other. This desirable complementary relationship exists only when one crop or soil-management practice concurrent with it provides nutrient or conditions required by the other crop. In this circumstance, grasses and legumes may complement grains or row crops by furnishing nitrogen, controlling erosion and pests, and improving soil structure to such an extent that greater production is achieved. The reverse can also occur; in certain prairie soils, continuous growing of deep-rooted legumes depletes soil moisture, and subsequent forage yield is improved by frequent plowing of the sod and planting of corn. In high-rainfall or irrigated areas, forage stands deteriorate from winter killing, disease, or grazing, to a point where a year of grain in the rotation allows an improved stand of forage later. Fallow (idle) land is complementary to wheat and other small grains in subhumid areas such as the Great Plains of the United States; such rotation is quite beneficial to wheat yield. Complementary relationships between crops can be terminated by the application of the physical law of diminishing returns, however, and give way to competition.

Both long-range and short-range profits motivate the farmer as cropping systems are examined in relationship to soil erosion. Excessive loss of soil to streams, rivers, and reservoirs is unacceptable to public policy as well as economically damaging to the farmer, and crop rotations that promote erosion are minimized. Soil losses are least from fields in continuous sod and most from continuous row crops. If row crops are grown in rotation with sod, the erosive susceptibility of row crops is reduced over a period of time. Peanuts (groundnuts), potatoes, tobacco, cotton, sugar beets, and some vegetables, and similar row crops that require frequent cultivation (intertillage) and leave minimal post-harvest residue are most likely to permit serious erosion. Less erosive are row crops such as corn (maize), sugarcane, and grain sorghum, which require less cultivation and leave more residue. Small grains such as wheat, oats, barley, and rye usually permit less erosion than the row crops. Among sod crops, grasses or grass–legume mixtures are less erosive than pure stands of legumes such as alfalfa. Fortunately, cropping systems that tend to control soil erosion usually tend also to give better yields than systems that promote excessive erosion. This results from increased availability of water to the plants and increased amounts of nutrients, which in erosive systems are washed away and lost.

Monoculture

The practice of growing the same crop each year on a given acreage, monoculture, has not been generally successful in the past, because nonlegume crops usually exhaust the nitrogen in the soil, with a resulting reduction in yields; this is particularly true in humid regions. The advent of low-cost nitrogen fertilizers, however, has induced reconsideration of the possible advantages of monoculture. These advantages can best be discussed in terms of a hypothetical general farm where it may be desirable to produce several different kinds of crops: the question to be answered is whether monoculture can do better than rotational systems in producing these crops while still maintaining productivity.

Advantages of Monoculture

First, if different kinds of soil exist on the farm, a monoculture system may permit each crop to be grown on the soil best suited to it. Forage crops, for example, could be confined to steep land to minimize erosion; intertilled crops could be planted on the better soils with gentle slopes. Wet areas could be used continuously for crops not requiring early-spring field operations, while dry soils could be used for drought-resistant crops such as sorghums or winter small grains.

Second, the fertility level of the soil can be adjusted to fit one crop more precisely than it can be adjusted to fit all the crops in a rotation.

Third, where continuous cropping is practiced and perennial forage crops are used, regular reseedings are avoided. This is an advantage, because each seeding is accompanied by the possibility of failure.

Fourth, systems based on monoculture usually offer greater flexibility in planning the system to meet year to year changes in the need for various crops. Part of the acreage can be shifted from one crop to another without upsetting the total farm cropping plan.

Disadvantages of Monoculture

On the other hand, requirements for successful monoculture are more demanding of management skill than are sod-based rotations. The entire nitrogen need of nonlegume crops must be met by purchased fertilizers or by use of manure. Closer attention to soil erosion is necessary, except for perennial sod. Soil-structure problems can become severe where intertilled crops are grown continuously. In monoculture, the farmer is completely dependent on chemical insecticides, disease-resistant plant varieties, soil fumigation, and similar methods of controlling insects and diseases that are usually controlled by crop rotation.

Thus, the choices of cropping systems that maintain good productivity, minimize soil losses, and are in harmony with demand and desired business organization are not easily made. The growing use of systems analysis will undoubtedly aid in rational selection among the bewildering array of possibilities.

Crop Protection

Crops are vulnerable to attack, damage, and competition. Insects, plant disease, nematodes, rodents, weeds, and air pollution are among the many enemies that can reduce crop yields and deny man the use of some of his farm-stored crops.

Insects, for example, can destroy a crop in a relatively short time. Control measures for many years have engaged the attention of farmer and scientist, yet full success has not been achieved, and the battle continues. The problem is further complicated by the fact that control measures not only kill unwanted insects, but also may harm honey bees as well as the parasites and predators that destroy insect pests.

At least 10,000 species of insects are classed as unwanted. Of these, several hundred species are particularly destructive and require some degree of control. They destroy food as well as the forage, pasture, and grain needed to produce livestock; and, in addition, they carry and transmit many diseases of plants and animals.

Chemical Control of Insects

Insecticides generally are effective, cheap, and safe if handled correctly; the good derived from them, however, can be partly offset by adverse effects. Chlorinated hydrocarbon insecticides such as DDT, for example, may leave residues toxic to beneficial insects, fish, and other wildlife; the insecticides may be found in meat and milk, or they may persist in the soil. Another problem is that some species of insects build up resistance to chlorinated hydrocarbon, organic phosphate, and carbamate insecticides. These disadvantages can be overcome only by persistent search for new and safer insecticides accompanied by wide use of nonchemical insect control.

A wide range of organophosphate and carbamate materials is now available. These can be applied to avoid most of the problems related to residues. Malathion and carbaryl, for example, are used to control insects in areas where persistent materials might appear later in meat or milk and can also be applied in areas where fish and wildlife might be affected. Those two chemicals offer a broad range of toxicity to insect pests. Unlike chlorinated hydrocarbons, they can be applied up to within a day or so of harvest without harm to many crops; they are dangerous, however, to those who apply them and must be handled with care.

Some insecticides are effective in very small amounts. This fact has stimulated development of ultralow-volume technology, where special equipment permits dispersal of low volumes of undiluted chemicals, which offers cost advantages as well as drastic reduction of the chemicals in the environment.

For example, six to 16 ounces (170 to 450 grams) per acre of Malathion may be effective against grasshoppers, boll weevil, cereal-leaf beetle (*Oulema melanopus*), mosquitoes, and the beet leafhopper (*Circulifer tenellus*). Formulation of chemicals in granules rather than sprays offers some advantages in use and applications; among others, it reduces the amount needed and also lessens the chance of adverse effects on beneficial insects and wildlife.

Certain insects that attack cotton, vegetables, and forage crops may be controlled by chemicals absorbed by the plant. Called systemics, they are placed with the seed at planting time. The chemical is taken up by the plant, and insects die when they attempt to feed on the leaf or stem. Beneficial insects that do not feed on the plant remain unharmed.

Nonchemical Control of Insects

Mechanical and Cultural Controls

Light traps that give off radiation that attracts insects have been under test for many years. They have been somewhat successful in controlling the codling moth (*Carpocapsa pomonella*) and the tobacco hornworm (*Protoparce sexta*).

Use of reflective aluminum strips, placed like a mulch in vegetable fields, has reduced or prevented aphid attack and thus protected cucumbers, squash, and watermelons from mosaic diseases. This technique may supplant insecticides, which frequently do not kill aphids quickly enough to prevent crop losses from virus transmitted by them.

For stored products, heat or cold can control many insects that frequent such places. Also, changing the proportions of oxygen, nitrogen, and carbon dioxide in the storage atmosphere can provide control.

It has been discovered that, if adult Indian-meal moths (*Plodia interpunctella*) were exposed to certain wavelengths of sound during the egg-laying period, their reproduction was reduced by 75 percent. The sound waves had a similar effect on flour beetles (*Tribolium* species). Further development is needed, but this method offers potential as a nonchemical control. Other types of physical energy can also kill insects. Light waves, high-frequency electric fields, high-intensity radio frequencies, and gamma radiation have been investigated; some offer promise.

Certain cultural practices can prevent or reduce insect crop damage. These include destruction of crop residues, deep plowing, crop rotation, use of fertilizers, strip-cropping, irrigation, and scheduled planting operations. Such practices are useful but cannot be relied upon entirely to eliminate severe infestations.

Biological Controls

The question of using biological controls has always been of considerable public interest. The control agents include parasites, predators, diseases, protozoa, and nematodes that attack the insect pests. Biological controls cannot replace insecticides entirely, because nature provides for survival of both beneficial and destructive insects. Before the population of a parasite or predator can expand, a high population of the host species must also be present. Sometimes the control agents are far outnumbered by the pest insect. Parasites and predators have furnished good control of the Japanese beetle (*Popillia japonica*), European corn borer (*Pyrausta nubilalis*), alfalfa aphid (*Therioaphis maculata*), alfalfa weevil (*Hypera postica*), and several others.

Microbial agents can be used for control. There exist about 1,100 viruses, bacteria, fungi, protozoa, rickettsiae, and nematodes that parasitize insects. Many pathogens are specific to a particular insect but are harmless to man and domestic animals. It is a possibility that insect pathogens can be produced, packaged, distributed, and applied in much the same way as insecticides.

The ideal solution to insect-control problems is to plant crop varieties that are resistant to attack. The only difficulty is that such varieties are not universally available, and development entails a very long process.

Sterilization of male insects by gamma radiation and their release into a population of wild insects is a promising approach. It has proved successful in control of screw-worms and fruit flies, replacing chemicals in some areas. Chemical attractants, which lure insects into contact with small amounts of insecticide or a sterilant, also offer much promise.

All aspects of insect control considered, it is possible that "integrated control," coordinated employment of more than one method, may be the answer. Combining resistant varieties with a systemic insecticide that leaves the parasites and predators unharmed, for example, has been successful in combatting the spotted alfalfa aphid in California. Preliminary reduction of heavy infestation by chemical spray combined with bait, followed by the sterile-insect technique, provides another example of integrated control. Use of sex attractant in light traps, plus special management of postharvest residues, has controlled the tobacco hornworm. Other examples might be cited, but the principal value of such control methods lies in using less insecticide and thus contributing to maintenance of a good environment.

Control of Plant Diseases and Nematodes

Insects, of course, are not the only agents hazardous to crops. Plant diseases and the microscopic worms called nematodes have the potential of creating wholesale destruction of crops, especially those grown in regions of wide weather fluctuation. In fact, these plant pests sometimes limit the kinds and varieties of crops that can be grown. The damage they cause may sometimes be mistaken for that caused by unfavourable weather. Epidemics may destroy crops completely.

As with insects, control of plant diseases and nematodes covers a broad spectrum of measures: use of chemicals, resistant varieties, quarantine, forecasting and warning, cultural practices, heat treatment, and others. Furthermore, most plant virus diseases are transmitted by insect carriers, so control of insects is linked to control of disease.

Nematodes and plant disease can at times be controlled fairly well by crop rotation, deep plowing, and burning of stubble and debris that remain after harvest. Though burning destroys aboveground organisms and permits economical control by chemicals, it contributes to air pollution and destroys organic matter. In another technique, propane-gas flame is applied to living plants as well as stubble to kill disease spores. A virus disease of sugarcane is controlled by heating diseased cuttings in hot-air ovens. Stem rot disease of peanuts (groundnuts) can be controlled by plowing under dead plant debris or by planting the seed on a raised bed followed by application of a pre-emergence weed killer.

Successful control of epidemic plant disease may depend on prompt application of chemicals before the disease outbreak. Many governments operate plant-disease forecasting and warning services for farmers. The service is based primarily on analysis of temperature, rainfall, humidity, and dew—all factors that can create conditions favourable to disease outbreaks.

Weed Control

Weed control is vital to agriculture, because weeds decrease yields, increase production costs, interfere with harvest, and lower product quality. Weeds also impede irrigation water-flow, interfere with pesticide application, and harbour disease organisms.

Early methods of weed control included mowing, flooding, cultivating, smothering, burning, and crop rotation. Though these methods are still important, other means are perhaps more typical today, particularly the use of herbicide (plant-killing) chemicals. Another technique is to introduce insects that attack only the unwanted plant and destroy it while leaving the crop plants unharmed.

The inadequacy of the cultural, mechanical, and biological control systems, however, stimulated the rapid development of chemical usage since World War II. Herbicides have had an impact on crop production, changing many cultural and mechanical agricultural operations.

Herbicides are formulated as wettable powders, granular materials, emulsions, and solutions. Any of them may be applied as a spot treatment, broadcast, placed in bands, or put directly on a specific plant part. When formulated as solutions or emulsions, the chemical is mixed with water or oil.

Spraying is the most common method, permitting extremely small amounts to be applied uniformly because of dilution. Sprays can be accurately directed underneath growing plants, and calibration and rate control are easier with spray machines than with granular applicators. Granular formulations have advantages under some conditions, however. The use of herbicides must be integrated into the overall farm program because the optimum date and application rate depend on the crop stage, the weed stage, weather conditions, and other factors.

Careful use of herbicides in farm production lowers cost, resulting in a more economical product for the consumer. Herbicides cut the costs of raising cotton, for example, by reducing labour requirements for weed control up to 60 percent. Herbicides replace hand labour in growing crops, labour that is no longer available in developed nations at costs the farmer can afford. Machines for chemical application are widely available.

When used as directed, herbicides are generally safe, not only for the operator but also for wildlife and livestock. The greatest difficulty lies in accidental injury to crop plants resulting from drift and from residues in the soil, particularly if residues enter water courses.

The future of chemical pesticides and herbicides is under debate by those who manufacture, sell, and use them and by those who are concerned about environmental quality. The value of an assured food and fibre supply at reasonable cost is undeniable, and chemicals contribute much toward this. These substances also cause undesirable effects upon the environment, however, and indeed can be toxic to a wide range of organisms. This fact will demand an increasing amount of care in using chemicals, perhaps enforced by law, along with increasing use of nonchemical control techniques.

Harvesting and Crop Processing

Harvesting Machinery

Harvesting machinery is generally classified by crop: reapers for cutting cereal grains and threshers for separating the seed from the plant. The more modern combine cuts, threshes, and cleans the grain in one operation. Corn (maize) harvesting is performed by mechanical corn pickers that snap the ears from the stalk so that only the grain and cobs are harvested. Corn shelling may be done mechanically in the field, after or with picking. Stripper-type cotton harvesters, which strip the entire plant of both open and unopened bolls, work best late in the season after frost has killed the green vegetative growth. Hay and forage machines include mowers, crushers, windrowers, field choppers, balers, and some machines that press the hay into wafers or pellets.

Grass, legumes, corn (maize), and other crops are often put into silos to keep them in a succulent and fermented state rather than stored dry as hay. To make silage, the crops must be cut up to permit tight packing in the silo, producing anaerobic fermentation and preventing formation of mold. Almost all silage crops are cut in the field with a forage harvester that cuts and chops the crop immediately or picks up and chops a windrow that has been cut and raked earlier.

Root crops are harvested with diggers and digger-pickers, which often pull up clods, stones, and vines with the crop. Though some machines carry workers who manually sort out extraneous material, this task is increasingly being performed mechanically. Modern sugar-beet harvesters lift the whole root from the ground, clean the earth from it, and deliver it to a bin or wagon. Sometimes the beet tops are removed before harvest of the roots and used for cattle feed. Peanuts (groundnuts) are lifted, vines and all, and allowed to dry before removal of the pods.

Tobacco-harvesting aids may be classified in three principal ways, according to the harvesting and curing methods used, which depend on the type of tobacco and its use. Flue-cured tobacco, a large plant that may stand three to four feet (90 to 120 centimetres) high, is harvested with machines that carry several workers who ride the lower platforms of the machines, cut the leaves, and place them on conveyor belts, where the leaves are tied mechanically or by hand. Burley tobacco has usually been harvested by workers using a machete-type knife. After cutting, the large end of the stalk is fixed onto the sharpened end of a stick, which—when loaded with a number of stalks—is hung by hand in a tobacco barn for curing. Researchers are attempting to mechanize the cutting, impaling, and hanging of burley tobacco. Little has been done, however, toward the mechanization of the harvesting of the small aromatic tobacco leaves, which are grown in the shade, picked by hand, tied with a string, then hung for curing.

Tree-crop harvesting is accomplished by hand or with mechanical shakers. Vegetable crops such as asparagus, lettuce, and cabbage are still harvested largely by hand, though scarcity and high cost of field labour has led to some mechanization in this area, notably with tomatoes.

Crop-processing Machinery

Machinery is widely used to prepare crops for convenient transportation, for safe storage, for the market, and for feeding to livestock. Advances in such machines have been rapid, particularly with new crops, increased yields, multiple-crop practices, and changing techniques.

In the most common method of crop drying, the crop, usually grain, is spread on floors or mats and stirred frequently while exposed to the sun. Such systems, though extremely common in the underdeveloped countries, are very slow and dependent on the weather. Forced-air-drying systems allow the farmer much more freedom in choosing grain varieties and harvest time. Fairly simple in operation, these systems have been gaining popularity in the tropics. Heat is often added to increase air temperatures during the drying period.

In a process called dryeration, wet corn (maize) is placed in a batch or continuous dryer. After losing 10 to 12 percent of its moisture, the hot corn is transferred to the dryeration cooling bin, in which it is tempered for six to 10 hours and then slowly cooled by ventilation for 10 hours. This process reduces kernel damage and increases dryer output. High moisture in stored hay not only causes rapid deterioration of its value as feed but often results in spontaneous combustion. When hay is first cut, it usually contains 70 percent or more moisture. It wilts and quickly dries to a moisture content of about 40 percent. At this stage, it can be dried to a safe storage condition, about 15 percent moisture, by blowing air through it, sometimes with supplemental heat.

Feed-processing mills, often referred to as feed grinders, are used principally for milling cereals for livestock feed, which aids digestion. The ground material is usually fairly coarse and at times may only be crushed. Modern mills frequently are designed to allow the farmer to grind the grain and to mix in various other ingredients in desired quantities. Other types of crop-processing machinery include machines that separate desirable seed from weed seed, stems and leaves, and dirt; grading machinery to classify seed by width, length, or thickness; fruit graders and separators; and cotton gins, which separate cotton seeds from the fibres.

Regional Variations in Technique

Dryland Farming

Dryland farming refers to production of crops without irrigation in regions where annual precipitation is less than 20 inches (500 millimetres). Where rainfall is less than 15 inches (400 millimetres) per year, winter wheat is the most favoured crop, although spring wheat is planted in some areas where severe winter killing may occur. (Grain sorghum is another crop grown in these areas.) Where some summer rainfall occurs, dry beans are an important crop. All dryland crop yield is mainly dependent on precipitation, but practices of soil management exert great influence on moisture availability and nutrient supply.

Where rainfall exceeds 15 inches (380 millimetres), the variety of crop possibilities is increased. In areas of favourable soils and moisture, seed alfalfa is grown, as is barley. Some grass seed may be grown, particularly crested wheat grass of various types.

Fallow System and Tillage Techniques

Dryland farming is made possible mainly by the fallow system of farming, a practice dating from ancient times. Basically, the term fallow refers to land that is plowed and tilled but left unseeded during a growing season. The practice of alternating wheat and fallow assumes that by clean cultivation the moisture received during the fallow period is stored for use during the crop season. Available soil nitrogen increases and weeds are controlled during the fallow period. One risk lies in the exposure of soil while fallow, leaving it susceptible to wind and water erosion. Modern power machinery has tended to reduce this risk.

Procedures and kinds of tillage that are comparatively new have proved effective in controlling erosion and improving water intake. Moldboard and disk plows are being replaced with chisels, sweeps, and other tools that stir and loosen the soil but leave the straw on the surface. Where the amount of straw or residue remaining from the previous crop is not excessive, this trashy fallow system works well, and tillage implements are designed to increase its effectiveness.

Contour tillage helps to prevent excessive runoff on moderate slopes. Broad terraces can aid in such moisture conservation. Steeper slopes are planted to permanent cover.

Compacted zones at a depth of five to eight inches (13 to 20 centimetres) can be caused by tillage. As such zones interfere with storage of moisture, they can be controlled by growing deep-rooted alfalfa at intervals, or the compacted zone can be broken by fall tillage with chisels or sweeps set to a depth just below the zone of compaction. Such deep tillage will result in reduced runoff and deeper moisture penetration.

When using power machinery in dryland farming, the timing of operations is important. The soil is broken in the fall or early spring before weeds or volunteer grain can deplete the moisture. Use of a rod weeder or similar equipment during fallow can control the weeds. Planting is timed to occur during the short period in fall or spring when temperature and moisture are favourable.

Fertilizer Use

Fertilizer is an important component of dryland technology. For example, 20 pounds per acre (22 kilograms per hectare) of nitrogen are recommended where rainfall is less than 13 inches (330 millimetres), ranging up to 60 pounds per acre (67 kilograms per hectare) where more rain is available; those figures refer to the production of wheat, but they are applicable to other dryland-farming areas. Where average annual precipitation is less than 12 inches (300 millimetres), the use of nitrogen is limited to years where moisture outlook is exceptionally favourable. Nitrogen fertilizer can be applied either in fall or spring. Band placement or broadcast techniques are utilized. Good results are obtained from broadcast spring application of nitrate fertilizer, and fall application of ammonia has also been successful. Local climates and rainfall patterns also determine choice of fertilizer and time of application.

Crops and Planting Methods

Alfalfa grown for seed on drylands is planted in rows, usually two to three feet (60 to 90 centimetres) apart; cultivation between rows is required during the first year. Alfalfa is also grown for forage where favourable. This practice builds nitrogen and organic matter, while improving soil structure. These legumes can be rotated with wheat if rain is between 16 and 18 inches (400 and 450 millimetres) and will increase the yield of wheat.

Other crops, such as cotton, peanuts (groundnuts), and grain sorghum, can be grown successfully in dryland agriculture. Where those crops are produced on sandy soils, special techniques are necessary to reduce soil blowing and drifting. Cotton and peanuts do not produce sufficient crop residues for protection from wind erosion, while sorghum does. For this reason, many farmers in such areas use various row combinations of cotton or peanuts with grain sorghum; two rows of grain sorghum and four to eight rows of peanuts in alternating strips is a popular technique. Another is to use a two-year rotation of cotton and grain sorghum, in which two rows are cropped and two rows are fallow. These systems not only afford protection from wind erosion but also promote effective use of soil moisture.

Tropical Farming

The area of the world bounded roughly on the north by the Tropic of Cancer and on the south by the Tropic of Capricorn, a vast land that embraces large parts of Latin America, Africa, India, Australia, and Southeast Asia, contains climates less favourable to agriculture and human settlement than those of the temperate zones.

Within this Equator-centred area occur the climates known as tropical, which are characterized by two general types: warm and wet, and warm with partly deficient rainfall. In either, the total precipitation is usually quite heavy, which leaches the tropical soils of nutrients. The area also has high temperatures with little variation the year round. The combination of high temperature and high rainfall causes organic matter to decompose quickly, leaving the soil deficient in humus. Vegetation flourishes in the tropics, along with weeds, insects, and disease organisms. Important climatic variations occur, depending upon land elevation.

Tropical crops include coconut, palm oil, rice, sugar, pineapple, sisal, cocoa, tea, coffee, jute, rubber, pepper, banana, and many others. In certain highland tropical areas, however, the crops common to temperate-climate agriculture can also be grown. The amount of tropical land well-suited to agriculture, however, is limited.

Plant-pest Problems

The abundance of plant pests in the tropics, including weeds and disease, makes agriculture successful mainly in the plantation system, where needed control measures can be financed. The alternative is to move from deteriorated land to newer fields; this practice of shifting agriculture has also been common, because tropical soils lose their productive capacity so rapidly.

The practice probably cannot be continued indefinitely, however, because of increasing population pressure.

The largest quantities of commercial tropical products originate in plantations, where skilled management is combined with sufficient capital to provide mechanized equipment.

This is particularly true in the production of coffee, cocoa, rubber, coconut, banana, pineapple, sugarcane, and others. Much rice is produced in the Asian tropics and Indonesia, however, on small farms with intensive hand labour and simple tools, where the prime mover is likely to be the ox or the water buffalo, not the tractor.

Water Management

Drainage, irrigation, and other special techniques of water management are important in tropical agriculture. An example is the cultivation of rice and sugarcane in the fertile coastal areas of Guyana. Originally through private enterprise and later by government efforts, large coastal areas were "empoldered" (diked) to keep back the sea in front and floods from the rivers in the rear. With a mean annual rainfall of 90 inches (2,300 millimetres), drainage is a critical factor; in fact, the system cannot discharge all possible floodwater, and so the crops must tolerate occasional drowning. With gravity drainage effective only at low tide, the drainage gates are opened on the ebbing tide and closed on the rising tide. Great difficulty is encountered in keeping the outlets unclogged by the heavy sediment discharge. Since rain does not always fall when it is needed, many fields are irrigated. Most of the rice soil is specially tilled after plowing in order to create a better seedbed under the water, using tractors operating in water four to six inches (10 to 15 centimetres) deep. After this special tillage, the seeds are broadcast in one to two inches (2.5 to five centimetres) of water. Though maintenance and operation of such an intricate water-control system are not simple, Guyana rice production has been doubled through its use.

Mechanical Problems

Mechanization faces many obstacles before wide adoption is possible in tropical regions. Difficult soils, stones, stumps, abundant labour, resistance from farmers, lack of incentives, lack of skills, lack of capital, low wages, high cost of machines, lack of dealer service, fragmented land ownership, all contribute to slow development of mechanization. Tropical soils differ markedly from those in the countries that manufacture land-preparation machinery, making adaptation of new design necessary. The encountering of stones, wood, trash, and termite mounds causes machines to break down. Depressing climatic conditions reduce the performance of the machine operators. Tropical farm regions are notoriously irregular or mountainous, impeding intensive machine culture. The best soils in Brazil require special erosion controls, reducing the potential for large-scale mechanization. One of the greatest overall impediments to mechanization is the fear that unemployment might result from it, a failure to understand that economic development and higher living standards depend partly on increasing the productivity of labour.

As an example of the problems encountered in mechanizing tropical crops, the harvesting experience of a large sugarcane plantation in

Trinidad is illuminating. On flatland of some 30,000 acres (12,000 hectares), the cane is grown on heavy clay soil in a climate with 50 inches (1,300 millimetres) of rain during the seven-month wet season and 10 inches during the five-month dry season. By 1960 the rising wage rate made harvest mechanization imperative. First, the traditional "bed" system, which functioned to remove floodwater, was changed to ridge planting; this made it possible for machines to operate and was a remarkable change in itself. Then it was decided to harvest with the cane combine, which tops, cuts, chops, and loads the chopped cane into transport vehicles. Although the combine is complicated and requires considerable power, it was deemed better than mechanical half-measures.

By 1969 the combines, however, were harvesting only 12.8 percent of the flatland crop, indicating that mechanization was far from complete. Three factors were responsible: first, the cane combines required extensive maintenance plus very expensive replacement parts. Second, it was difficult to mobilize a transport system to receive the output of the combines with any degree of economy. Third, the social problem of displaced workers had to be considered. The combines increased labour productivity sixfold over hand harvesting; thus, their introduction had to be slowed until surplus workers could be accommodated elsewhere. The limited success of this mechanization project indicates how complicated such a process really is.

Taking the largest view of possibilities for improving tropical agriculture, the most promising inputs of technology are improved crop varieties and increased use of fertilizers.

Other Specialized Techniques

Hydroponics

The term hydroponics denotes soilless culture of plants. The possibilities of this technique have received considerable attention in recent years. In hydroponics, an outgrowth of laboratory techniques long used by scientists, plants are grown with their roots immersed in a water solution containing necessary minerals or rooted in a sand medium kept moistened by such a solution. Soilless culture of plants is similar in principle but larger in scale. A typical hydroponics technique has plants supported in a bed of peat, wood fibre, or similar material, on a wire screen with the roots dipping into the solution below. Aeration of the solution is provided. In another method, the plants are rooted in a medium of sand or gravel contained in a shallow tank into which the solution is pumped at intervals by automatic control. Between pumpings, the solution drains slowly down into a reservoir tank. Hydroponic techniques are practiced on a small scale both out-of-doors and in greenhouses.

Of the elements known to be necessary for plant growth, carbon, oxygen, and hydrogen are obtained by the plant from atmospheric gases or from soil water. The others are all obtained as mineral salts from the soil. The elements absorbed as salts—iron, manganese, boron, copper, zinc, and molybdenum—are required in minute quantities and are called the micronutrients. The principal elements that must be provided as dissolved salts in hydroponic techniques are nitrogen, phosphorus, sulfur, potassium, calcium, and magnesium. Numerous solutions have been devised to fulfill these requirements.

Crop yields of some plants can be obtained fully equal to those obtained on fertile soils. Wide-scale crop production by hydroponics, however, would be economic only for certain intensive types of agriculture or under special conditions. Some greenhouse crops, both vegetables and flowers, are grown by this method. In regions having no soil or extremely infertile soil but with favourable climate, hydroponic techniques have been very useful; for example, on some of the coral islands of the Pacific.

Greenhouses

The greenhouse is typically a structure whose roof and sides are transparent or translucent, permitting a sufficient quality and quantity of solar radiation to enter the structure for photosynthesis. It allows the growing of crops independently of the outside climate, since its interior temperature and humidity can be controlled. Greenhouses vary in size and complexity from small home or hobby structures to large commercial units covering an acre or more of land. An even smaller greenhouse might be termed the hot bed, a glass-topped box containing fermenting organic matter; the fermentation process yields heat, allowing the gardener to start plants from seed in early spring for later transplanting.

The basic construction of a greenhouse consists of a light but sturdy frame capable of resisting winds and other loads. Conventional foundations usually support vertical walls; the roof may be gabled, trussed, or arched. The conventional greenhouse is fitted with glass panes, but plastic-film or fibre-glass panels often supplant glass.

Maintenance of temperature within the greenhouse is difficult because of fluctuating outside conditions. When the sun shines brightly, little heat is needed, and the heating system must be controlled in some way to prevent injury to the crop. Hot water, steam, electric cable, or warm-air furnaces provide the heat, which is usually controlled by thermostat. Temperatures in greenhouses are regulated to suit the crop. Typical ranges are from 40° F (4° C) for lettuce, violets, carnations, and sweet peas to 70° F (21° C) for cucumbers, tomatoes, and orchids.

Cooling is often required during summer days in warm climates. Ventilation is the simplest technique, reducing inside temperature to near that of the outdoors. Additional cooling by refrigeration may be required; in dry regions, the evaporative cooler is efficient and also increases the relative humidity within the structure. Another form of environmental control consists of adding extra carbon dioxide to the air if the crop requires it for extra photosynthetic efficiency.

The commercial-greenhouse operator usually grows vegetables or ornamental plants. Such production makes more demands on the grower, because he must assume many of the tasks normally handled by nature in the open fields. He must regulate the temperature, ventilate, adjust the amount of entering sunlight, provide soil moisture, fertilize, and even facilitate pollination. During the off-season, the structure must be cleaned and fumigated, its soil restructured, and mechanical equipment checked. Mechanization of greenhouse operations has lagged far behind the pattern of agriculture in general. Disease is a particularly serious hazard in greenhouse farming, requiring constant attention and use of chemicals.

The factor of Weather

Weather Information

The interaction of weather and living systems is a basic aspect of agriculture. Although great strides in technology have resulted in massive production increases and improved quality, weather remains an important limiting factor. Though man is not yet able to change the weather, except on a very small scale, he is capable of adjusting agricultural practices to fit the climate.

Thus, weather information is of utmost importance when combined with other factors, such as knowledge of crop or livestock response to weather factors; the farmer's capability to act on alternative decisions based on available weather information; existence of two-way communication by which specific weather forecasts and allied information can be requested and distributed; and the climatic probability of occurrence of influential weather elements and the ability of the meteorologist to predict their occurrence.

Other Weather-research Benefits

Apart from the many applications of weather forecasting to current problems, meteorological research may benefit agriculture in at least three other ways:

(1) improved planning of widescale land usage depends partly on detailed knowledge of plant-climate interactions; radiation, evapotranspiration, diurnal temperature range, water balance, and other parameters are measured and analyzed before a plan realizing maximum economic benefit for a given area is prepared;

(2) agronomic experiments are combined with climatological documentation to obtain the greatest scientific and technological return;

(3) problems of irrigation, row spacing, timing of fertilizer application, variety selection, and transplanting can best be solved with the aid of climatic environmental data; cultural practices related to artificial modification of microclimates should be based on research knowledge rather than personal judgment.

Observing Climatic Elements

The climatic elements the observation of which is valuable for agricultural purposes can be approached on an idealized threefold scale:

(1) microscale observations of small areas for research designed to elucidate basic physical processes;

(2) mesoscale climatic networks designed for practicing farmers to improve their operations; and

(3) macroscale regional networks intended for weather forecasting and for gathering basic climatic data.

Macroscale stations can be further divided into first-order and second-order stations, the number and type of observations different for each. Micrometeorology demands the most elaborate array of measuring devices, while a second-order macroscale station requires the least; in fact, the latter station will measure only five elements: air temperature, rain, snow, humidity, and surface wind.

A first-order macrostation will be equipped to measure 16 elements: global radiation, sunshine hours, clouds, net radiation, air temperature, soil temperature, rain, snow, hail, dew, fog, humidity, pan evaporation, pressure, upper air wind, and surface wind. Mesoscale measurements include 10 elements and microscale 27 (three of which are derived from others).

The World Meteorological Organization and the various national weather services are concerned with establishment and improvement of macroscale regional climatic stations, both first-class and second-class. Spaced at least 10 miles (16 kilometres) apart, their value for daily agricultural operations is limited, but they are useful for long-range planning and forecasting. Most parts of North America, Europe, and Australia have adequate networks of these stations, but wide gaps exist in the tropics, polar regions, and arid lands.

The Degree Day

One weather characteristic of agricultural value is the degree day. This concept holds that the growth of a plant is dependent on the total amount of heat to which it is subjected during its lifetime, accumulated as degree days. Common practice is to use 50° F (10° C) as a base. Thus, if the mean daily temperature for a particular day is 60° F (16° C), then 10 degree days are accumulated for that day on the Fahrenheit scale. The total number of growing degree days required for maturity varies with crop variety as well as plant species. Also, the minimum threshold temperature (the temperature below which the plant is damaged or unable to grow) varies with plants; *e.g.,* 40° F (4° C) for peas, 50° F (10° C) for corn (maize), and 55° F (13° C) for citrus fruits. Where studies have established the number of degree days required for maturity of a given crop, the planting dates can be scheduled for orderly harvest and processing. The system is helpful in selecting crop varieties appropriate to different geographical areas; it also has value in scheduling spray programs and predicting insect emergence.

The growing-degree-day concept has certain weaknesses:

(1) it assumes that the relationship between growth and temperature is linear (actually it is not);

(2) it makes no allowance for changing threshold temperatures with advancing crop development;

(3) too much weight is given to temperatures above 80° F (27° C), which may be detrimental; and

(4) no account is taken of the diurnal temperature range, which is often more significant than the mean daily value.

Weather Effects

The essence of the weather–agriculture interaction for the farmer lies in wise adaptation of operations to the local climate and in techniques for manipulating or modifying the local environment (microclimate) to minimize weather stresses on plants and animals.

Many of these techniques have been practiced for centuries: seeding and cultivation, irrigation, frost protection, animal shelters, windbreaks, and others are methods of altering the microclimate. The climatic factors and their relation to plant growth in terms of protective techniques are important.

Solar Radiation

Solar radiation is the ultimate source for all physical and biological processes of the earth. Agriculture itself is a strategy for exploitation of solar energy, made possible by water and nutrients. During daytime hours, solar radiation is delivered both directly and by diffused sky reflection. The incoming radiation that is not reflected by the surface or reradiated to outer space is the net radiation, which is the energy available for maintaining the earth's surface temperature. At night the net radiation is negative; that is, energy is lost to outer space by long-wave radiation, and none is gained. The net radiation balance varies widely throughout the world, setting limits on basic agricultural possibilities.

Photosynthesis

Photosynthesis is the process by which higher plants manufacture dry matter through the aid of chlorophyll pigment, which uses solar energy to produce carbohydrates out of water and carbon dioxide. The overall efficiency of this critical process is somewhat low, and its mechanics are extremely complex. It is related to light intensity, wavelength, temperature, carbon dioxide concentration in the air, and the respiration rate of the plant. The distribution of solar energy within the plant community is affected by the leaf canopy's density, height, and capacity to transmit the energy; these therefore affect photosynthesis. The leaf-foliage density is characterized by the leaf-area index, the total leaf area of a plant over a given area of land. The optimum leaf-area index will vary between summer and winter and between temperate and tropical regions, but it represents a key factor in the search for better crop management based on improved photosynthesis. The efficiency of radiation utilization by field crops has been measured, showing that an ordinary crop converts less than 1 percent of available solar energy into organic matter.

Photoperiodism

Photoperiodism is another attribute of plants that may be changed or manipulated in the microclimate. The length of a day is a photoperiod, and the responses of the plant development to a photoperiod are called photoperiodism. Response to the photoperiod is different for different plants; long-day plants flower only under day lengths longer than 14 hours; in short-day plants, flowering is induced by photoperiods of less than 10 hours; day-neutral plants form buds under any period of illumination. There are exceptions and variations in photoperiodic response; also, it is argued that the truly critical factor is actually the amount of exposure to darkness rather than to daylight. Temperature is intimately related to photoperiodism, tending to modify reactions to daylength. Photoperiodism is one determining factor in natural distribution of plants throughout the world.

The phenomenon has many practical applications. Selection of a plant or a variety for a given locality requires knowledge of its interaction with the photoclimate. Artificial illumination is used to control flowering seasons and to increase production of greenhouse crops. In plant breeding, such stimulation of flowering has greatly reduced the time span from germination to maturity, shortening the time necessary to develop new varieties. In sowing field crops, photoperiodism can be used to select the date of sowing to produce optimum harvest size. Crop yield is reduced both by planting in a season that will cause plants to flower early and by planting at a time that will cause very late flowering. In Sri Lanka (formerly Ceylon), certain rice varieties with a vegetative period of five to six months may extend their life to more than a year when planted in the wrong season, causing almost complete loss of yield. Cowpeas in Nigeria will flower early and produce many seeds only when planted in daylengths of 12 hours or less.

Weather Conditions and Controls

Temperature

Regardless of how favourable light and moisture conditions may be, plant growth ceases when the air and leaf temperature drops below a certain minimum or exceeds a certain maximum value. Between these limits, there is an optimum temperature at which growth proceeds with greatest rapidity. These three temperature points are the cardinal temperatures for a given plant; the cardinal temperatures are known for most plant species, at least approximately. Cool-season crops (oats, rye, wheat, and barley) have low cardinal temperatures: minimum 32° to 41° F (0° to 5° C), optimum 77° to 88° F (25° tï 31° C), and maximum 88° tï 99° F (31° to 37° C). For hot-season crops, such as melons and sorghum, the span of cardinal temperatures is much higher. The cardinal temperatures may vary with stage of development. For example, cold treatment near 32° F (0° C) of germinated seeds before sowing can transform winter rye into the spring type; such treatment, called vernalization, has practical application in cold-climate plants. The range of diurnal temperature variation is also important; the best net photosynthesis is related to a large diurnal temperature range, or high daytime and low nighttime temperatures. Knowledge of the difference between leaf and air temperatures aids farmers in adopting protective measures. In middle and high latitudes, frost often occurs before the air temperature drops to freezing; in summer, heat injury to plants might be much more serious than that suggested by the air temperature alone. Because of this factor, farmers in Taiwan shade the pineapple fruit to prevent heat damage.

Soil temperature sometimes is of greater ecological significance to plant life than air temperature. Germination of seed, root function, rate of plant growth, and occurrence and severity of plant diseases all are affected by soil temperature. Since an unfavourable soil temperature during the growing season can retard or ruin a crop, techniques have been developed for modifying the temperature. The two most important methods are (1) regulation of the energy exchange and (2) altering the thermal properties of the ground. Incoming energy can be regulated by an insulation layer on or near the ground surface, such as paper, straw, plastic, or trees; the outgoing radiation can be reduced by insulation materials or by generating smoke or fog in the air. Thermal properties of the ground can be modified by cultivation or irrigation, increasing the soil's ability to absorb radiation, or by varying the rate of evaporation. Mulching is a common technique for soil temperature control. Carbon black or white material can change the soil's ability to absorb radiation. In the Soviet Union, for example, it was reported that 100 pounds of coal dust per acre (112 kilograms per hectare) caused a one-month advance in the maturity date of cotton.

Frost

Another aspect of temperature control is frost protection. Likelihood of damage from freezing temperature depends upon the plant species, the season, the manner of temperature change, the physiological state of the plant, and other factors. Orchards can be located so as to minimize the chances of frost damage.

Two types of frost are recognized: (1) radiation frost, which occurs on clear nights with little or no wind when the outgoing radiation is excessive and the air temperature is not necessarily at the freezing point, and (2) wind, or advection, frost, which occurs at any time, day or night, regardless of cloud cover, when wind moves air in from cold regions. Both types may occur simultaneously. Most frost-protection techniques can raise the temperature only a few degrees, while some are effective only against radiation frost.

Heating is probably the best known and most effective frost-protection measure. It is most effective on nights with a strong temperature inversion, a condition in which the air temperature increases markedly from the ground up to as high as 40 or 50 feet (12 or 15 metres). The depth of air to be heated is thus rather shallow, and the area over which a given temperature rise can be produced increases linearly with the strength of the inversion. Lacking a temperature inversion, heaters protect by radiating heat to the plants and the ground surface, and by emitting a layer of humid smoke that reduces the net outgoing loss from the ground.

In general, a large number of small heaters is most effective; large heaters set up convection currents that break up the warm ceiling and draw in cold air. For radiation-frost protection, the heaters are placed in "view" of the plants or trees, but for advective frost the heavier concentration is placed along the upwind border. Common fuels for the heaters include oil, coal, briquettes, and wood. Oil is most effective, because it can be ignited rapidly and extinguished easily. Heating is a costly technique; a few growers who tried it in England soon gave up the practice, and, even in places such as California, heating is becoming less common and is mostly restricted to a few high-value crops such as citrus fruits.

The wind machine is popular for frost protection; although it affords less reliable results, its operating cost is much lower than that for heaters. These machines, which are like fans or propellers, break up the nocturnal temperature inversion by mechanically mixing the air, returning heat to the ground that was lifted during the day. The stronger the temperature inversion, the more effective is the wind machine, which is ineffective, however, against a daytime freeze or cold soil. Even under the best circumstances, ground-surface temperatures will rise very little; therefore, some operators install both heaters and wind machines, using the latter for strong-inversion nights and the former for wind-frost protection.

Flooding and sprinkling with water prevent excessive ground cooling by increasing the heat conductivity and heat capacity of the soil and releasing latent heat of fusion, or the heat given off when the water freezes. The temperature of the plant will not fall below the freezing point so long as the change of state from water to ice is taking place. Flooding has the disadvantage of retarding increase in soil warmth during the day; thus, it can be used effectively for only one or two nights. Sprinkling creates water particles in the air that reduce outgoing radiation, but plant temperature declines immediately on cessation of sprinkling, and the ice formation may cause damage to the crop. In general, successful protection by flooding and sprinkling demands much skill and judgment from the operator.

Brushing is a frost-protection technique in which shields of paper or aluminum foil are set up to reduce radiation loss to the sky; it has been used with fair success for tomato culture in California.

Massachusetts cranberry growers add a thin layer of sand to the soil periodically. The sandy surface warms up easily and cools slowly by radiation; it also reduces evaporation of its low water content. Sanding can raise the temperature of loam, clay, and organic soils, thus diminishing frost hazard. Windbreaks can also function as frost protection by reducing inflow of cold air and by shielding plants from the total night sky.

Spraying of harmless foams or gels on plants threatened by frost is a technique under investigation. The trapped air in the foam serves as insulative protection, while the foam can be designed to dissolve after any desired time interval. The technique has been explored for use on strawberries and other low-growing crops.

Irrigation

Irrigation is probably the most common form of agricultural microclimatic control practiced by man. Also important are efforts to correct deficiencies in precipitation, the deficiencies that lead farmers to irrigate.

Rainmaking

Attempts to increase the amount of precipitation from clouds by seeding them with salt or silver iodide have been made for nearly three decades. Both aircraft and ground generators have been employed, but the techniques are typically beyond the means of an individual farmer. Results suggest that cloud modification is entirely possible, but the proof of increased rainfall at a level of statistical significance is a difficult problem. Success has been greatest under atmospheric conditions where natural rainfall is most probable. The prospect of modifying winter clouds to increase snowfall in mountain areas appears to be somewhat more promising, however.

Most cloud-seeding efforts are expended in regions where precipitation is only marginal for agriculture. It is commonly assumed that at least 20 inches (500 millimetres) of rain per year, fairly well distributed, is required to maintain a stable farming community. Unfortunately, the years of large deficiencies in such areas are those with only limited opportunity for cloud seeding. Some observers believe that weather modification to increase precipitation may yet become practical and economically feasible; the legal, ethical, and ecological problems raised by the prospect will not be easily solved, however.

Humidity

The value of high humidity in the greenhouse is well known, but knowledge of humidity–plant interaction under field conditions is comparatively slight. Other things being equal, the evapotranspiration rate decreases with increasing humidity; thus, rate of water use is higher at low levels of humidity. The benefits of irrigation are apparently greater when the humidity is high, which simply means that the efficiency of water use increases with humidity.

Wind

Wind affects plant growth in at least three significant ways: transpiration, carbon dioxide intake, and mechanical breakage. Transpiration (the loss of water mainly through the stomata of leaves) increases with wind speed, but the effect varies greatly among plant species; also, the effect is related to temperature and humidity of the air. In arid climates, dry and hot winds often cause rapid, harmful wilting. In winter, with frozen soil, the damaging effect of increased transpiration resulting from wind can be serious because the lost water cannot be readily replaced.

By contrast, increasing wind promotes carbon dioxide intake within limits; this benefits the rate of photosynthesis. The effects of mechanical wind damage vary from species to species; some show a definite decrease in dry matter production with increasing wind, while others (usually short plants) are unaffected. Because of the long-recognized need, shelterbelts, massive plantings of trees that change the energy and moisture balance of the crop, are positioned to protect crops and to increase yields. A shelterbelt perpendicular to the prevailing wind reduces velocity on both sides.

A medium-thick shelterbelt can reduce wind velocity by more than 10 percent to a distance of 20 times the tree height on the leeward side and three times the tree height to the windward. The length of the shelterbelt should be at least equal to that of the field to be protected. The sheltered area will suffer much less soil erosion and mechanical damage than unprotected areas. Other microclimatic effects of shelterbelts include:

(1) small daytime temperature increases and nighttime decreases;

(2) the occurrence of radiation frost in the leeside may be promoted;

(3) rate of evaporation in the sheltered area is decreased, depending on wind velocity;

(4) snow accumulates near the shelterbelt, causing increased moisture storage in dry farming.

The overall effect of a shelterbelt is complicated but probably beneficial. There is much evidence that they increase efficiency of water use not only in subhumid and semi-arid regions but also in true deserts where oasis-type irrigation is practiced. The response to shelterbelts, however, depends on the species. Crops of low response to wind protection are the drought-hardy small grains and maize grown under dry farming conditions. Rice and forage crops such as alfalfa, lupine, and clover are moderately responsive. Crops that benefit most from wind protection are garden crops, such as lentils, potatoes, tomatoes, cucumbers, beets, strawberries, watermelons, deciduous and citrus fruits, and other tender crops, such as tobacco and tea. Some authorities assert that in strong wind areas shelterbelts will produce an average 20 percent yield increase, which is net gain of 15 percent when allowance is made for the land occupied by the belts themselves. Trees can be grown almost anywhere, even in the desert; tall plants such as corn (maize), sorghums, or even elephant grass can also be employed in arid regions by including them in the irrigation schedule. It would appear that windbreaks are among the most practical means of beneficial weather modification in agriculture.

Chapter 2

Electrical Technology in Agriculture

Electricity

Electricity is the set of physical phenomena associated with the presence and flow of electric charge. Electricity gives a wide variety of well-known effects, such as lightning, static electricity, electromagnetic induction and electrical current. In addition, electricity permits the creation and reception of electromagnetic radiation such as radio waves.

In electricity, charges produce electromagnetic fields which act on other charges. Electricity occurs due to several types of physics:

- electric charge: a property of some subatomic particles, which determines their electromagnetic interactions. Electrically charged matter is influenced by, and produces, electromagnetic fields.
- electric field: an especially simple type of electromagnetic field produced by an electric charge even when it is not moving (i.e., there is no electric current). The electric field produces a force on other charges in its vicinity.
- electric potential: the capacity of an electric field to do work on an electric charge, typically measured in volts.
- electric current: a movement or flow of electrically charged particles, typically measured in amperes.
- electromagnets: Moving charges produce a magnetic field. Electrical currents generate magnetic fields, and changing magnetic fields generate electrical currents.

In electrical engineering, electricity is used for:

- electric power where electric current is used to energise equipment;
- electronics which deals with electrical circuits that involve active electrical components such as vacuum tubes, transistors, diodes and integrated circuits, and associated passive interconnection technologies.

Electrical phenomena have been studied since antiquity, though progress in theoretical understanding remained slow until the seventeenth and eighteenth centuries. Even then, practical applications for electricity were few, and it would not be until the late nineteenth century that engineers were able to put it to industrial and residential use. The rapid expansion in electrical technology at this time transformed industry and society. Electricity's extraordinary versatility means it can be put to an almost limitless set of applications which include transport, heating, lighting, communications, and computation. Electrical power is now the backbone of modern industrial society.

History

Long before any knowledge of electricity existed people were aware of shocks from electric fish. Ancient Egyptian texts dating from 2750 BC referred to these fish as the "Thunderer of the Nile", and described them as the "protectors" of all other fish. Electric fish were again reported millennia later by ancient Greek, Roman and Arabic naturalists and physicians. Several ancient writers, such as Pliny the Elder and Scribonius Largus, attested to the numbing effect of electric shocks delivered by catfish and torpedo rays, and knew that such shocks could travel along conducting objects. Patients suffering from ailments such as gout or headache were directed to touch electric fish in the hope that the powerful jolt might cure them. Possibly the earliest and nearest approach to the discovery of the identity of lightning, and electricity from any other source, is to be attributed to the Arabs, who before the 15th century had the Arabic word for lightning (*raad*) applied to the electric ray.

Ancient cultures around the Mediterranean knew that certain objects, such as rods of amber, could be rubbed with cat's fur to attract light objects like feathers. Thales of Miletus made a series of observations on static electricity around 600 BC, from which he believed that friction rendered amber magnetic, in contrast to minerals such as magnetite, which needed no rubbing. Thales was incorrect in believing the attraction was due to a magnetic effect, but later science would prove a link between magnetism and electricity. According to a controversial theory, the Parthians may have had knowledge of electroplating, based on the 1936 discovery of the Baghdad Battery, which resembles a galvanic cell, though it is uncertain whether the artifact was electrical in nature.

Electricity would remain little more than an intellectual curiosity for millennia until 1600, when the English scientist William Gilbert made a careful study of electricity and magnetism, distinguishing the lodestone effect from static electricity produced by rubbing amber. He coined the New Latin word *electricus* ("of amber" or "like amber", from $ëåêôñïí, *elektron*, the Greek word for "amber") to refer to the property of attracting small objects after being rubbed. This association gave rise to the English words "electric" and "electricity", which made their first appearance in print in Thomas Browne's *Pseudodoxia Epidemica* of 1646.

Further work was conducted by Otto von Guericke, Robert Boyle, Stephen Gray and C. F. du Fay. In the 18th century, Benjamin Franklin conducted extensive research in electricity, selling his possessions to fund his work. In June 1752 he is reputed to have attached a metal key to the bottom of a dampened kite string and flown the kite in a storm-threatened sky. A succession of sparks jumping from the key to the back of his hand showed that lightning was indeed electrical in nature. He also explained the apparently paradoxical behaviour of the Leyden jar as a device for storing large amounts of electrical charge in terms of electricity consisting of both positive and negative charges.

In 1791, Luigi Galvani published his discovery of bioelectricity, demonstrating that electricity was the medium by which nerve cells passed signals to the muscles. Alessandro Volta's battery, or voltaic pile, of 1800, made from alternating layers of zinc and copper, provided scientists with a more reliable source of electrical energy than the electrostatic machines previously used. The recognition of electromagnetism, the unity of electric and magnetic phenomena, is due to Hans Christian Ørsted and André-Marie Ampère in 1819-1820; Michael Faraday invented the electric motor in 1821, and Georg Ohm mathematically analysed the electrical circuit in 1827. Electricity and magnetism (and light) were definitively linked by James Clerk Maxwell, in particular in his "On Physical Lines of Force" in 1861 and 1862.

While the early 19th century had seen rapid progress in electrical science, the late 19th century would see the greatest progress in electrical engineering. Through such people as Alexander Graham Bell, Ottó Bláthy, Thomas Edison, Galileo Ferraris, Oliver Heaviside, Ányos

Jedlik, Lord Kelvin, Sir Charles Parsons, Ernst Werner von Siemens, Joseph Swan, Nikola Tesla and George Westinghouse, electricity turned from a scientific curiosity into an essential tool for modern life, becoming a driving force of the Second Industrial Revolution.

In 1887, Heinrich Hertz discovered that electrodes illuminated with ultraviolet light create electric sparks more easily. In 1905 Albert Einstein published a paper that explained experimental data from the photoelectric effect as being the result of light energy being carried in discrete quantized packets, energising electrons. This discovery led to the quantum revolution. Einstein was awarded the Nobel Prize in 1921 for "his discovery of the law of the photoelectric effect". The photoelectric effect is also employed in photocells such as can be found in solar panels and this is frequently used to make electricity commercially.

The first solid-state device was the "cat's whisker" detector, first used in 1930s radio receivers. A whisker-like wire is placed lightly in contact with a solid crystal (such as a germanium crystal) in order to detect a radio signal by the contact junction effect. In a solid-state component, the current is confined to solid elements and compounds engineered specifically to switch and amplify it. Current flow can be understood in two forms: as negatively charged electrons, and as positively charged electron deficiencies called holes. These charges and holes are understood in terms of quantum physics. The building material is most often a crystalline semiconductor.

The solid-state device came into its own with the invention of the transistor in 1947. Common solid-state devices include transistors, microprocessor chips, and RAM. A specialized type of RAM called flash RAM is used in flash drives and more recently, solid state drives to replace mechanically rotating magnetic disc hard drives. Solid state devices became prevalent in the 1950s and the 1960s, during the transition from vacuum tubes to semiconductor diodes, transistors, integrated circuit (IC) and the light-emitting diode (LED).

Concepts

Electric Charge

The presence of charge gives rise to an electrostatic force: charges exert a force on each other, an effect that was known, though not understood, in antiquity. A lightweight ball suspended from a string can be charged by touching it with a glass rod that has itself been charged by rubbing with a cloth. If a similar ball is charged by the same glass rod, it is found to repel the first: the charge acts to force the two balls apart. Two balls that are charged with a rubbed amber rod also repel each other. However, if one ball is charged by the glass rod, and the other by an amber rod, the two balls are found to attract each other. These phenomena were investigated in the late eighteenth century by Charles-Augustin de Coulomb, who deduced that charge manifests itself in two opposing forms. This discovery led to the well-known axiom: *like-charged objects repel and opposite-charged objects attract.*

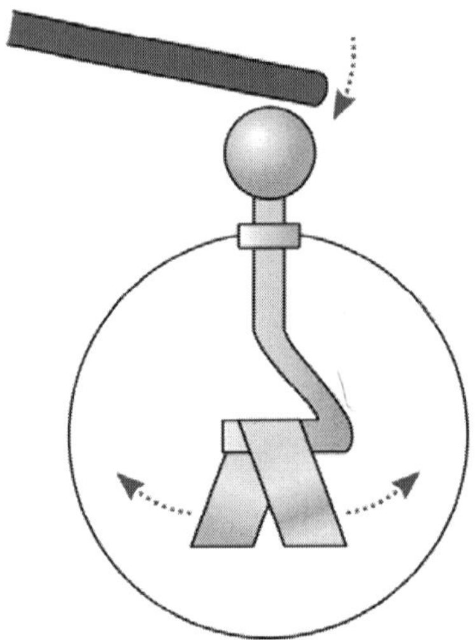

Figure: *Charge on a gold-leaf electroscope causes the leaves to visibly repel each other*

The force acts on the charged particles themselves, hence charge has a tendency to spread itself as evenly as possible over a conducting surface. The magnitude of the electromagnetic force, whether attractive or repulsive, is given by Coulomb's law, which relates the force to the product of the charges and has an inverse-square relation to the distance between them. The electromagnetic force is very strong, second only in strength to the strong interaction, but unlike that force it operates over all distances. In comparison with the much weaker gravitational force, the electromagnetic force pushing two electrons apart is 10^{42} times that of the gravitational attraction pulling them together.

Study has shown that the origin of charge is from certain types of subatomic particles which have the property of electric charge. Electric charge gives rise to and interacts with the electromagnetic force, one of the four fundamental forces of nature. The most familiar carriers of electrical charge are the electron and proton. Experiment has shown charge to be a conserved quantity, that is, the net charge within an isolated system will always remain constant regardless of any changes taking place within that system. Within the system, charge may be transferred between bodies, either by direct contact, or by passing along a conducting material, such as a wire. The informal term static electricity refers to the net presence (or 'imbalance') of charge on a body, usually caused when dissimilar materials are rubbed together, transferring charge from one to the other.

The charge on electrons and protons is opposite in sign, hence an amount of charge may be expressed as being either negative or positive. By convention, the charge carried by electrons is deemed negative, and that by protons positive, a custom that originated with the work of Benjamin Franklin. The amount of charge is usually given the symbol Q and expressed in coulombs; each electron carries the same charge of approximately -1.6022×10^{-19} coulomb. The proton has a charge that is equal and opposite, and thus $+1.6022\times10^{-19}$ coulomb. Charge is possessed not just by matter, but also by antimatter, each antiparticle bearing an equal and opposite charge to its corresponding particle.

Charge can be measured by a number of means, an early instrument being the gold-leaf electroscope, which although still in use for classroom demonstrations, has been superseded by the electronic electrometre.

Electric Current

The movement of electric charge is known as an electric current, the intensity of which is usually measured in amperes. Current can consist of any moving charged particles; most commonly these are electrons, but any charge in motion constitutes a current.

By historical convention, a positive current is defined as having the same direction of flow as any positive charge it contains, or to flow from the most positive part of a circuit to the most negative part. Current defined in this manner is called conventional current. The motion of negatively charged electrons around an electric circuit, one of the most familiar forms of current, is thus deemed positive in the *opposite* direction to that of the electrons. However, depending on the conditions, an electric current can consist of a flow of charged particles in either direction, or even in both directions at once. The positive-to-negative convention is widely used to simplify this situation.

The process by which electric current passes through a material is termed electrical conduction, and its nature varies with that of the charged particles and the material through which they are travelling. Examples of electric currents include metallic conduction, where electrons flow through a conductor such as metal, and electrolysis, where ions (charged atoms) flow through liquids, or through plasmas such as electrical sparks. While the particles themselves can move quite slowly, sometimes with an average drift velocity only fractions of a millimetre per second, the electric field that drives them itself propagates at close to the speed of light, enabling electrical signals to pass rapidly along wires.

Current causes several observable effects, which historically were the means of recognising its presence. That water could be decomposed by the current from a voltaic pile was discovered by Nicholson and Carlisle in 1800, a process now known as electrolysis. Their work was greatly expanded upon by Michael Faraday in 1833. Current through a resistance causes localised heating, an effect James Prescott Joule studied mathematically in 1840. One of the most important discoveries relating to current was made accidentally by Hans Christian Ørsted in 1820, when, while preparing a lecture, he witnessed the current in a wire disturbing the needle of a magnetic compass. He had discovered electromagnetism, a fundamental interaction between electricity and magnetics. The level of electromagnetic emissions generated by electric arcing is high enough to produce electromagnetic interference, which can be detrimental to the workings of adjacent equipment.

In engineering or household applications, current is often described as being either direct current (DC) or alternating current (AC). These terms refer to how the current varies in time. Direct current, as produced by example from a battery and required by most electronic devices, is a unidirectional flow from the positive part of a circuit to the negative. If, as is most common, this flow is carried by electrons, they will be travelling in the opposite direction. Alternating current is any current that reverses direction repeatedly; almost always this takes the form of a sine wave. Alternating current thus pulses back and forth within a conductor without the charge moving any net distance over time. The time-averaged value of an alternating current is zero, but it delivers energy in first one direction, and then the reverse. Alternating current is affected by electrical properties that are not observed under steady state direct current, such as inductance and capacitance. These properties however can become important when circuitry is subjected to transients, such as when first energised.

Electric Field

The concept of the electric field was introduced by Michael Faraday. An electric field is created by a charged body in the space that surrounds it, and results in a force exerted on any other charges placed within the field. The electric field acts between two charges in a similar manner to the way that the gravitational field acts between two masses, and like it, extends towards infinity and shows an inverse square relationship with distance. However, there is an important difference. Gravity always acts in attraction, drawing two masses together, while the electric field can result in either attraction or repulsion. Since large bodies such as planets generally carry no net charge, the electric field at a distance is usually zero. Thus gravity is the dominant force at distance in the universe, despite being much weaker.

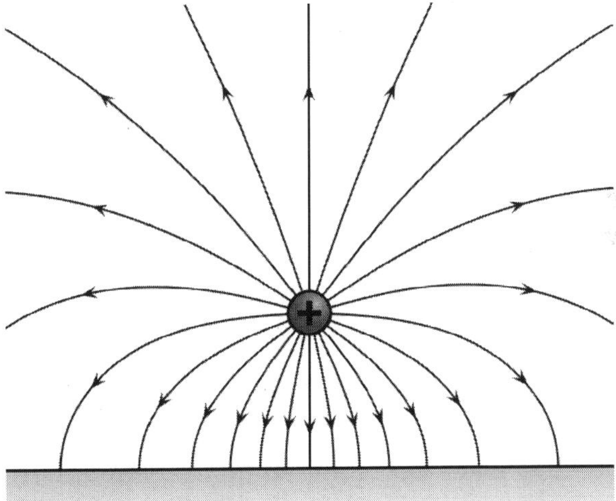

Figure: *Field lines emanating from a positive charge above a plane conductor*

An electric field generally varies in space, and its strength at any one point is defined as the force (per unit charge) that would be felt by a stationary, negligible charge if placed at that point. The conceptual charge, termed a 'test charge', must be vanishingly small to prevent its own electric field disturbing the main field and must also be stationary to prevent the effect of magnetic fields. As the electric field is defined in terms of force, and force is a vector, so it follows that an electric field is also a vector, having both magnitude and direction. Specifically, it is a vector field.

The study of electric fields created by stationary charges is called electrostatics. The field may be visualised by a set of imaginary lines whose direction at any point is the same as that of the field. This concept was introduced by Faraday, whose term 'lines of force' still sometimes sees use. The field lines are the paths that a point positive charge would seek to make as it was forced to move within the field; they are however an imaginary concept with no physical existence, and the field permeates all the intervening space between the lines. Field lines emanating from stationary charges have several key properties: first, that they originate at positive charges and terminate at negative charges; second, that they must enter any good conductor at right angles, and third, that they may never cross nor close in on themselves.

A hollow conducting body carries all its charge on its outer surface. The field is therefore zero at all places inside the body. This is the operating principal of the Faraday cage, a conducting metal shell which isolates its interior from outside electrical effects.

The principles of electrostatics are important when designing items of high-voltage equipment. There is a finite limit to the electric field strength that may be withstood by any medium. Beyond this point, electrical breakdown occurs and an electric arc causes flashover between the charged parts. Air, for example, tends to arc across small gaps at electric field strengths which exceed 30 kV per centimetre. Over larger gaps, its breakdown strength is weaker, perhaps 1 kV per centimetre. The most visible natural occurrence of this is lightning, caused when charge becomes separated in the clouds by rising columns of air, and raises the electric field in the air to greater than it can withstand. The voltage of a large lightning cloud may be as high as 100 MV and have discharge energies as great as 250 kWh.

The field strength is greatly affected by nearby conducting objects, and it is particularly intense when it is forced to curve around sharply pointed objects. This principle is exploited in the lightning conductor, the sharp spike of which acts to encourage the lightning stroke to develop there, rather than to the building it serves to protect

Electric Potential

The concept of electric potential is closely linked to that of the electric field. A small charge placed within an electric field experiences a force, and to have brought that charge to that point against the force requires work. The electric potential at any point is defined as the energy required to bring a unit test charge from an infinite distance slowly to that point. It is usually measured in volts, and one volt is the potential for which one joule of work must be expended to bring a charge of one coulomb from infinity. This definition of potential, while formal, has little practical application, and a more useful concept is that of electric potential difference, and is the energy required to move a unit charge between two specified points. An electric field has the special property that it is *conservative*, which means that the path taken by the test charge is irrelevant: all paths between two specified points expend the same energy, and thus a unique value for potential difference may be stated. The volt is so strongly identified as the unit of choice for measurement and description of electric potential difference that the term voltage sees greater everyday usage.

For practical purposes, it is useful to define a common reference point to which potentials may be expressed and compared. While this could be at infinity, a much more useful reference is the Earth itself, which is assumed to be at the same potential everywhere. This reference point naturally takes the name earth or ground. Earth is assumed to be an infinite source of equal amounts of positive and negative charge, and is therefore electrically uncharged—and unchargeable.

Electric potential is a scalar quantity, that is, it has only magnitude and not direction. It may be viewed as analogous to height: just as a released object will fall through a difference in heights caused by a gravitational field, so a charge will 'fall' across the voltage caused by an electric field. As relief maps show contour lines marking points of equal height, a set of lines marking points of equal potential (known as equipotentials) may be drawn around an electrostatically charged object. The equipotentials cross all lines of force at right angles. They must also lie parallel to a conductor's surface, otherwise this would produce a force that will move the charge carriers to even the potential of the surface.

The electric field was formally defined as the force exerted per unit charge, but the concept of potential allows for a more useful and equivalent definition: the electric field is the local gradient of the electric potential. Usually expressed in volts per metre, the vector direction of the field is the line of greatest slope of potential, and where the equipotentials lie closest together.

Electromagnets

Orsted's discovery in 1821 that a magnetic field existed around all sides of a wire carrying an electric current indicated that there was a direct relationship between electricity and magnetism. Moreover, the interaction seemed different from gravitational and electrostatic forces, the two forces of nature then known. The force on the compass needle did not direct it to or away from the current-carrying wire, but acted at right angles to it. Ørsted's slightly obscure words were that "the electric conflict acts in a revolving manner." The force also depended on the direction of the current, for if the flow was reversed, then the force did too.

Orsted did not fully understand his discovery, but he observed the effect was reciprocal: a current exerts a force on a magnet, and a magnetic field exerts a force on a current. The phenomenon was further investigated by Ampère, who discovered that two parallel current-carrying wires exerted a force upon each other: two wires conducting currents in the same direction are attracted to each other, while wires containing currents in opposite directions are forced apart.

The interaction is mediated by the magnetic field each current produces and forms the basis for the international definition of the ampere.

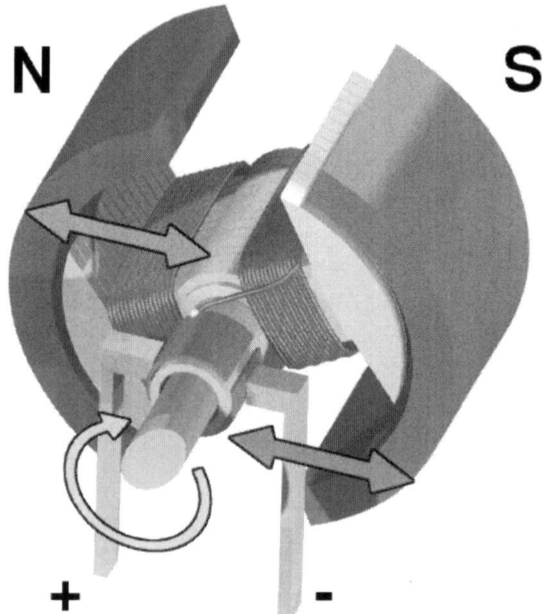

Figure: *The electric motor exploits an important effect of electromagnetism: a current through a magnetic field experiences a force at right angles to both the field and current*

This relationship between magnetic fields and currents is extremely important, for it led to Michael Faraday's invention of the electric motor in 1821. Faraday's homopolar motor consisted of a permanent magnet sitting in a pool of mercury.

A current was allowed through a wire suspended from a pivot above the magnet and dipped into the mercury. The magnet exerted a tangential force on the wire, making it circle around the magnet for as long as the current was maintained.

Experimentation by Faraday in 1831 revealed that a wire moving perpendicular to a magnetic field developed a potential difference between its ends. Further analysis of this process, known as electromagnetic induction, enabled him to state the principle, now known as Faraday's law of induction, that the potential difference induced in a closed circuit is proportional to the rate of change of magnetic flux through the loop. Exploitation of this discovery enabled him to invent the first electrical generator in 1831, in which he converted the mechanical energy of a rotating copper disc to electrical energy. Faraday's disc was inefficient and of no use as a practical generator, but it showed the possibility of generating electric power using magnetism, a possibility that would be taken up by those that followed on from his work.

Electrochemistry

The ability of chemical reactions to produce electricity, and conversely the ability of electricity to drive chemical reactions has a wide array of uses.

Electrochemistry has always been an important part of electricity. From the initial invention of the Voltaic pile, electrochemical cells have evolved into the many different types of batteries, electroplating and electrolysis cells. Aluminium is produced in vast quantities this way, and many portable devices are electrically powered using rechargeable cells.

Electric Power

Figure: *Electric power is transmitted on overhead lines like these, and also on underground high voltage cables.*

Electric power is the rate at which electric energy is transferred by an electric circuit. The SI unit of power is the watt, one joule per second.

Electric power is usually produced by electric generators, but can also be supplied by sources such as electric batteries. It is generally supplied to businesses and homes by the electric power industry through an electric power grid. Electric power is usually sold by the kilowatt hour (3.6 MJ) which is the product of power in kilowatts multiplied by running time in hours. Electric utilities measure power using an electricity metre, which keeps a running total of the electric energy delivered to a customer.

Definition

Electric power, like mechanical power, is the rate of doing work, measured in watts, and represented by the letter P. The term *wattage* is used colloquially to mean "electric power in watts." The electric power in watts produced by an electric current I consisting of a charge of Q coulombs every t seconds passing through an electric potential (voltage) difference of V is

$$P = \text{work done per unit time} = \frac{VQ}{t} = VI$$

where

Q is electric charge in coulombs

t is time in seconds

I is electric current in amperes

V is electric potential or voltage in volts

An alternate way to derive this formula is to note that voltage is defined as $V = dW / dQ$, the amount of work that a unit charge (one coulomb) does when it moves between the two terminals, and the current is defined as $I = dQ / dt$, the number of coulombs flowing per second, so

$$P = \frac{dW}{dt} \quad \text{(rate of work done per unit time)}$$

$$= \left(\frac{dW}{dQ}\right)\left(\frac{dQ}{dt}\right) \text{(work done per unit charge} \times \text{charge flowing per unit time)}$$

$= VI$ (voltage$\times$current)

Explanation

Electric power is transformed to other forms of power when electric charges move through an electric potential (voltage) difference, which occurs in electrical components in electric circuits. From the standpoint of electric power, components in an electric circuit can be divided into two categories:

- Passive devices or loads: When electric charges move through a potential difference from a high voltage to a low voltage, that is conventional current (positive charge) moves from the positive terminal to the negative, the potential energy of the charges is converted to kinetic energy, which performs work on the device. Devices in which this occurs are called *passive* devices or *loads*; they consume electric power from the circuit, converting it to other forms such as mechanical work, heat, light, etc. Examples are electrical appliances, such as light bulbs, electric motors, and electric heaters. In alternating current (AC) circuits the direction of the current and voltage periodically reverses, but the instantaneous current is always moving from the high potential to the low potential side.
- Active devices or power sources: If the charges are forced by an outside force to move through the device in the direction from a lower electric potential to a higher, so positive charge moves from the negative to the positive terminal, work is being done *on* the charges, so energy is being converted to electric potential energy from some other type of energy, such as mechanical energy or chemical energy. Devices in which this occurs are called *active* devices or *power sources*; sources of electric current, such as electric generators and batteries.

Some devices can be either a source or a load, depending on the voltage or current through them. For example, a rechargeable battery acts as a source when it provides power to a circuit, but as a load when it is connected to a battery charger and is being recharged.

Passive sign Convention

Since electric power can flow in either direction, either into or out of a component, a convention is needed for which direction represents positive power flow. Electric power flowing *out* of a circuit *into* a component is arbitrarily defined to have a positive sign, while power flowing *into* a circuit from a component is defined to have a negative sign. Thus passive components have positive power consumption. This is called the *passive sign convention.*

Resistive Circuits

In the case of resistive (Ohmic, or linear) loads, Joule's law can be combined with Ohm's law ($V = I \cdot R$) to produce alternative expressions for the dissipated power:

$$P = I^2 R = \frac{V^2}{R},$$

where R is the electrical resistance.

Alternating Current

In alternating current circuits, energy storage elements such as inductance and capacitance may result in periodic reversals of the direction of energy flow. The portion of power flow that, averaged over a complete cycle of the AC waveform, results in net transfer of energy in one direction is known as real power (also referred to as active power). That portion of power flow due to stored energy, that returns to the source in each cycle, is known as reactive power. The real power P in watts consumed by a device is given by

$$P = \frac{1}{2} V_p I_p \cos\theta = V_{rms} I_{rms} \cos\theta$$

where

V_p is the peak voltage in volts

I_p is the peak current in amperes

V_{rms} is the root-mean-square voltage in volts

I_{rms} is the root-mean-square current in amperes

θ is the phase angle between the current and voltage sine waves

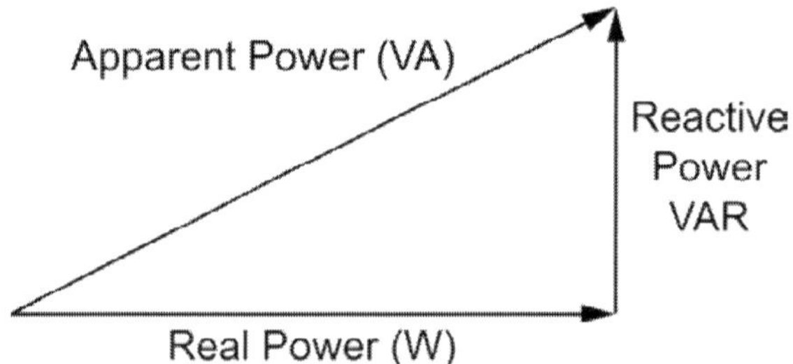

Figure: *Power triangle: The components of AC power*

The relationship between real power, reactive power and apparent power can be expressed by representing the quantities as vectors. Real power is represented as a horizontal vector and reactive power is represented as a vertical vector. The apparent power vector is the hypotenuse of a right triangle formed by connecting the real and reactive power vectors. This representation is often called the *power triangle*. Using the Pythagorean Theorem, the relationship among real, reactive and apparent power is:

$$(\text{apparent power})^2 = (\text{real power})^2 + (\text{reactive power})^2$$

Real and reactive powers can also be calculated directly from the apparent power, when the current and voltage are both sinusoids with a known phase angle θ between them:

$$(\text{real power}) = (\text{apparent power})\cos\theta$$

$$(\text{reactive power}) = (\text{apparent power})\sin\theta$$

The ratio of real power to apparent power is called power factor and is a number always between 0 and 1. Where the currents and voltages have non-sinusoidal forms, power factor is generalized to include the effects of distortion

Electromagnetic fields

Electrical energy flows wherever electric and magnetic fields exist together and fluctuate in the same place. The simplest example of this is in electrical circuits, as the preceding section showed.

In the general case, however, the simple equation $P = IV$ must be replaced by a more complex calculation, the integral of the cross-product of the electrical and magnetic field vectors over a specified area, thus:

$$P = \int_S (\mathrm{E} \times \mathrm{H}) \cdot \mathrm{dA}.$$

The result is a scalar since it is the *surface integral* of the *Poynting vector*.

Electric Power Supply

Electricity Generation

The fundamental principles of electricity generation were discovered during the 1820s and early 1830s by the British scientist Michael Faraday. His basic method is still used today: electricity is generated by the movement of a loop of wire, or disc of copper between the poles of a magnet.

For electric utilities, it is the first process in the delivery of electricity to consumers. The other processes, electricity transmission, distribution, and electrical power storage and recovery using pumped-storage methods are normally carried out by the electric power industry.

Electricity is most often generated at a power station by electromechanical generators, primarily driven by heat engines fueled by chemical combustion or nuclear fission but also by other means such as the kinetic energy of flowing water and wind.

There are many other technologies that can be and are used to generate electricity such as solar photovoltaics and geothermal power.

Battery Power

A battery is a device consisting of one or more electrochemical cells that convert stored chemical energy into electrical energy. Since the invention of the first battery (or "voltaic pile") in 1800 by Alessandro Volta and especially since the technically improved Daniell cell in 1836, batteries have become a common power source for many household and industrial applications. According to a 2005 estimate, the worldwide battery industry generates US$48 billion in sales each year, with 6% annual growth.

There are two types of batteries: primary batteries (disposable batteries), which are designed to be used once and discarded, and secondary batteries (rechargeable batteries), which are designed to be recharged and used multiple times. Batteries come in many sizes, from miniature cells used to power hearing aids and wristwatches to battery banks the size of rooms that provide standby power for telephone exchanges and computer data centres.

Electric Power Industry

The electric power industry provides the production and delivery of power, in sufficient quantities to areas that need electricity, through a grid connection. The grid distributes electrical energy to customers. Electric power is generated by central power stations or by distributed generation.

Many households and businesses need access to electricity, especially in developed nations, the demand being scarcer in developing nations. Demand for electricity is derived from the requirement for electricity in order to operate domestic appliances, office equipment, industrial machinery and provide sufficient energy for both domestic and commercial lighting, heating, cooking and industrial processes. Because of this aspect of the industry, it is viewed as a public utility as infrastructure.

Electromagnetic Wave

Faraday's and Ampère's work showed that a time-varying magnetic field acted as a source of an electric field, and a time-varying electric field was a source of a magnetic field. Thus, when either field is changing in time, then a field of the other is necessarily induced. Such a phenomenon has the properties of a wave, and is naturally referred to as an electromagnetic wave. Electromagnetic waves were analysed theoretically by James Clerk Maxwell in 1864. Maxwell developed a set of equations that could unambiguously describe the interrelationship between electric field, magnetic field, electric charge, and electric current. He could moreover prove that such a wave would necessarily travel at the speed of light, and thus light itself was a form of electromagnetic radiation. Maxwell's Laws, which unify light, fields, and charge are one of the great milestones of theoretical physics.

Thus, the work of many researchers enabled the use of electronics to convert signals into high frequency oscillating currents, and via suitably shaped conductors, electricity permits the transmission and reception of these signals via radio waves over very long distances.

Production and Uses

Generation and Transmission

Thales' experiments with amber rods were the first studies into the production of electrical energy. While this method, now known as the triboelectric effect, can lift light objects and generate sparks, it is extremely inefficient. It was not until the invention of the voltaic pile in the eighteenth century that a viable source of electricity became available. The voltaic pile, and its modern descendant, the electrical battery, store energy chemically and make it available on demand in the form of electrical energy. The battery is a versatile and very common power source which is ideally suited to many applications, but its energy storage is finite, and once discharged it must be disposed of or recharged. For large electrical demands electrical energy must be generated and transmitted continuously over conductive transmission lines.

Electrical power is usually generated by electro-mechanical generators driven by steam produced from fossil fuel combustion, or the heat released from nuclear reactions; or from other sources such as kinetic energy extracted from wind or flowing water. The modern steam turbine invented by Sir Charles Parsons in 1884 today generates about 80 percent of the electric power in the world using a variety of heat sources.

Such generators bear no resemblance to Faraday's homopolar disc generator of 1831, but they still rely on his electromagnetic principle that a conductor linking a changing magnetic field induces a potential difference across its ends. The invention in the late nineteenth century of the transformer meant that electrical power could be transmitted more efficiently at a higher voltage but lower current. Efficient electrical transmission meant in turn that electricity could be generated at centralised power stations, where it benefited from economies of scale, and then be despatched relatively long distances to where it was needed.

Since electrical energy cannot easily be stored in quantities large enough to meet demands on a national scale, at all times exactly as much must be produced as is required. This requires electricity utilities to make careful predictions of their electrical loads, and maintain constant co-ordination with their power stations. A certain amount of generation must always be held in reserve to cushion an electrical grid against inevitable disturbances and losses.

Demand for electricity grows with great rapidity as a nation modernises and its economy develops. The United States showed a 12% increase in demand during each year of the first three decades of the twentieth century, a rate of growth that is now being experienced by emerging economies such as those of India or China. Historically, the growth rate for electricity demand has outstripped that for other forms of energy.

Environmental concerns with electricity generation have led to an increased focus on generation from renewable sources, in particular from wind and hydropower. While debate can be expected to continue over the environmental impact of different means of electricity production, its final form is relatively clean

Applications

Electricity is a very convenient way to transfer energy, and it has been adapted to a huge, and growing, number of uses. The invention of a practical incandescent light bulb in the 1870s led to lighting becoming one of the first publicly available applications of electrical power. Although electrification brought with it its own dangers, replacing the naked flames of gas lighting greatly reduced fire hazards within homes and factories. Public utilities were set up in many cities targeting the burgeoning market for electrical lighting.

The Joule heating effect employed in the light bulb also sees more direct use in electric heating. While this is versatile and controllable, it can be seen as wasteful, since most electrical generation has already required the production of heat at a power station. A number of countries, such as Denmark, have issued legislation restricting or banning the use of electric heating in new buildings. Electricity is however a highly practical energy source for refrigeration, with air conditioning representing a growing sector for electricity demand, the effects of which electricity utilities are increasingly obliged to accommodate.

Electricity is used within telecommunications, and indeed the electrical telegraph, demonstrated commercially in 1837 by Cooke and

Wheatstone, was one of its earliest applications. With the construction of first intercontinental, and then transatlantic, telegraph systems in the 1860s, electricity had enabled communications in minutes across the globe.

Optical fibre and satellite communication have taken a share of the market for communications systems, but electricity can be expected to remain an essential part of the process.

The effects of electromagnetism are most visibly employed in the electric motor, which provides a clean and efficient means of motive power. A stationary motor such as a winch is easily provided with a supply of power, but a motor that moves with its application, such as an electric vehicle, is obliged to either carry along a power source such as a battery, or to collect current from a sliding contact such as a pantograph.

Electronic devices make use of the transistor, perhaps one of the most important inventions of the twentieth century, and a fundamental building block of all modern circuitry. A modern integrated circuit may contain several billion miniaturised transistors in a region only a few centimetres square.

Electricity is also used to fuel public transportation, including electric buses and trains.

Ampère's Circuital Law

In classical electromagnetism, Ampère's circuital law, discovered by André-Marie Ampère in 1826, relates the integrated magnetic field around a closed loop to the electric current passing through the loop. James Clerk Maxwell derived it again using hydrodynamics in his 1861 paper *On Physical Lines of Force* and it is now one of the Maxwell equations, which form the basis of classical electromagnetism.

Ampère's Original Circuital Law

Ampère's law relates magnetic fields to electric currents that produce them. Ampère's law determines the magnetic field associated with a given current, or the current associated with a given magnetic field, provided that the electric field does not change over time. In its original form, Ampère's circuital law relates a magnetic field to its electric current source.

The law can be written in two forms, the "integral form" and the "differential form". The forms are equivalent, and related by the Kelvin Stokes theorem. It can also be written in terms of either the B or H magnetic fields. Again, the two forms are equivalent.

Ampère's circuital law is now known to be a correct law of physics in a magnetostatic situation: The system is static except possibly for continuous steady currents within closed loops. In all other cases the law is incorrect unless Maxwell's correction is included.

Integral form

In SI units (cgs units are later), the "integral form" of the original Ampère's circuital law is a line integral of the magnetic field around some closed curve *C* (arbitrary but must be closed). The curve *C* in turn bounds both a surface *S* which the electric current passes through (again arbitrary but not closed—since no three-dimensional volume is enclosed by *S*), and encloses the current. The mathematical statement of the law is a relation between the total amount of magnetic field around some path (line integral) due to the current which passes through that enclosed path (surface integral). It can be written in a number of forms.

In terms of total current, which includes both free and bound current, the line integral of the magnetic B-field (in tesla, T) around closed curve *C* is proportional to the total current I_{enc} passing through a surface *S* (enclosed by *C*):

$$\oint_C \mathrm{B} \cdot \mathrm{d}\ell = \mu_0 \iint_S \mathrm{J} \cdot \mathrm{dS} = \mu_0 I_{enc}$$

where J is the total current density (in ampere per square metre, Am^{-2}).

Alternatively in terms of free current, the line integral of the magnetic H-field (in ampere per metre, Am^{-1}) around closed curve *C* equals the free current $I_{f,\,enc}$ through a surface *S*:

$$\oint_C \mathrm{H} \cdot \mathrm{d}\ell = \iint_S \mathrm{J}_f \cdot \mathrm{dS} = I_{f,enc}$$

where J_f is the free current density only. Furthermore

- $\oint_C$ is the closed line integral around the closed curve *C*,
- $\iint_S$ denotes a 2d surface integral over *S* enclosed by *C*
- • is the vector dot product,
- $d\ell$ is an infinitesimal element (a differential) of the curve *C* (i.e. a vector with magnitude equal to the length of the infinitesimal line element, and direction given by the tangent to the curve *C*)
- dS is the vector area of an infinitesimal element of surface *S* (that is, a vector with magnitude equal to the area of the infinitesimal surface element, and direction normal to surface

S. The direction of the normal must correspond with the orientation of *C* by the right hand rule).

The B and H fields are related by the constitutive equation

$$B = \mu_0 H$$

where μ_0 is the magnetic constant.

There are a number of ambiguities in the above definitions that require clarification and a choice of convention.

1. First, three of these terms are associated with sign ambiguities: the line integral $\oint_C$ could go around the loop in either direction (clockwise or counterclockwise); the vector area dS could point in either of the two directions normal to the surface; and I_{enc} is the net current passing through the surface *S*, meaning the current passing through in one direction, minus the current in the other direction—but either direction could be chosen as positive. These ambiguities are resolved by the right-hand rule: With the palm of the right-hand toward the area of integration, and the index-finger pointing along the direction of line-integration, the outstretched thumb points in the direction that must be chosen for the vector area dS. Also the current passing in the same direction as dS must be counted as positive. The right hand grip rule can also be used to determine the signs.
2. Second, there are infinitely many possible surfaces *S* that have the curve *C* as their border. (Imagine a soap film on a wire loop, which can be deformed by moving the wire). Which of those surfaces is to be chosen? If the loop does not lie in a single plane, for example, there is no one obvious choice. The answer is that it does not matter; it can be proven that any surface with boundary *C* can be chosen.

Differential form

By the Stokes' theorem, this equation can also be written in a "differential form". Again, this equation only applies in the case where the electric field is constant in time, meaning the currents are steady (time-independent, else the magnetic field would change with time). In SI units, the equation states for total current:

$$\nabla \times B = \mu_0 J$$

and for free current

$$\nabla \times H = J_f$$

where $\nabla \times$ is the curl operator.

Note on free Current Versus Bound Current

The electric current that arises in the simplest textbook situations would be classified as "free current"—for example, the current that passes through a wire or battery. In contrast, "bound current" arises in the context of bulk materials that can be magnetized and/or polarized. (All materials can to some extent.)

When a material is magnetized (for example, by placing it in an external magnetic field), the electrons remain bound to their respective atoms, but behave as if they were orbiting the nucleus in a particular direction, creating a microscopic current.

When the currents from all these atoms are put together, they create the same effect as a macroscopic current, circulating perpetually around the magnetized object. This magnetization current J_M is one contribution to "bound current".

The other source of bound current is bound charge. When an electric field is applied, the positive and negative bound charges can separate over atomic distances in polarizable materials, and when the bound charges move, the polarization changes, creating another contribution to the "bound current", the polarization current J_P.

The total current density J due to free and bound charges is then:

$$J = J_f + J_M + J_P$$

with J_f the "free" or "conduction" current density.

All current is fundamentally the same, microscopically. Nevertheless, there are often practical reasons for wanting to treat bound current differently from free current. For example, the bound current usually originates over atomic dimensions, and one may wish to take advantage of a simpler theory intended for larger dimensions. The result is that the more microscopic Ampère's law, expressed in terms of B and the microscopic current (which includes free, magnetization and polarization currents), is sometimes put into the equivalent form below in terms of H and the free current only.

Shortcomings of the Original formulation of Ampère's Circuital Law

There are two important issues regarding Ampère's law that require closer scrutiny. First, there is an issue regarding the continuity equation for electrical charge. In vector calculus, the identity for the divergence of a curl states that a vector field's curl divergence must always be zero. Hence

$$\nabla \cdot (\nabla \times B) = 0$$

and so the original Ampère's law implies that

$$\nabla \cdot J = 0.$$

But in general

$$\nabla \cdot J = -\frac{\partial \rho}{\partial t}$$

which is non-zero for a time-varying charge density. An example occurs in a capacitor circuit where time-varying charge densities exist on the plates.

Second, there is an issue regarding the propagation of electromagnetic waves. For example, in free space, where

$$J = 0.$$

Ampère's law implies that

$$\nabla \times B = 0$$

but instead

$$\nabla \times B = \frac{1}{c^2}\frac{\partial E}{\partial t}.$$

To treat these situations, the contribution of displacement current must be added to the current term in Ampère's law.

James Clerk Maxwell conceived of displacement current as a polarization current in the dielectric vortex sea, which he used to model the magnetic field hydrodynamically and mechanically. He added this displacement current to Ampère's circuital law at equation (112) in his 1861 paper *On Physical Lines of Force* .

Displacement Current

In free space, the displacement current is related to the time rate of change of electric field.

In a dielectric the above contribution to displacement current is present too, but a major contribution to the displacement current is related to the polarization of the individual molecules of the dielectric material. Even though charges cannot flow freely in a dielectric, the charges in molecules can move a little under the influence of an electric field. The positive and negative charges in molecules separate under the applied field, causing an increase in the state of polarization, expressed as the polarization density P. A changing state of polarization is equivalent to a current.

Both contributions to the displacement current are combined by defining the displacement current as:

$$J_D = \frac{\partial}{\partial t} D(r, t),$$

where the electric displacement field is defined as:

$$D = \varepsilon_0 E + P = \varepsilon_0 \varepsilon_r E,$$

where ε_0 is the electric constant, ε_r the relative static permittivity, and P is the polarization density. Substituting this form for D in the expression for displacement current, it has two components:

$$J_D = \varepsilon_0 \frac{\partial E}{\partial t} + \frac{\partial P}{\partial t}.$$

The first term on the right hand side is present everywhere, even in a vacuum. It doesn't involve any actual movement of charge, but it nevertheless has an associated magnetic field, as if it were an actual current. Some authors apply the name *displacement current* to only this contribution.

The second term on the right hand side is the displacement current as originally conceived by Maxwell, associated with the polarization of the individual molecules of the dielectric material.

Maxwell's original explanation for displacement current focused upon the situation that occurs in dielectric media. In the modern post-aether era, the concept has been extended to apply to situations with no material media present, for example, to the vacuum between the plates of a charging vacuum capacitor.

The displacement current is justified today because it serves several requirements of an electromagnetic theory: correct prediction of magnetic fields in regions where no free current flows; prediction of wave propagation of electromagnetic fields; and conservation of electric charge in cases where charge density is time-varying.

Extending the Original Law: the Maxwell–Ampère Equation

Next Ampère's equation is extended by including the polarization current, thereby remedying the limited applicability of the original Ampère's circuital law.

Treating free charges separately from bound charges, Ampère's equation including Maxwell's correction in terms of the H-field is (the H-field is used because it includes the magnetization currents, so J_M does not appear explicitly):

$$\oint_C H \cdot d\ell = \iint_S \left(J_f + \frac{\partial}{\partial t} D \right) \cdot dS$$

(integral form), where H is the magnetic H field (also called "auxiliary magnetic field", "magnetic field intensity", or just "magnetic field"), D is the electric displacement field, and J_f is the enclosed conduction current or free current density. In differential form,

$$\nabla \times \mathrm{H} = \mathrm{J}_f + \frac{\partial}{\partial t}\mathrm{D}.$$

On the other hand, treating all charges on the same footing (disregarding whether they are bound or free charges), the generalized Ampère's equation, also called the Maxwell–Ampère equation, is in integral form:

$$\oint_C \mathrm{B} \cdot d\ell = \iint_S \left(\mu_0 \mathrm{J} + \mu_0 \epsilon_0 \frac{\partial}{\partial t}\mathrm{E} \right) \cdot d\mathrm{S}$$

In differential form,

$$\nabla \times \mathrm{B} = \left(\mu_0 \mathrm{J} + \mu_0 \epsilon_0 \frac{\partial}{\partial t}\mathrm{E} \right)$$

In both forms J includes magnetization current density as well as conduction and polarization current densities. That is, the current density on the right side of the Ampère–Maxwell equation is:

$$\mathrm{J}_f + \mathrm{J}_D + \mathrm{J}_M = \mathrm{J}_f + \mathrm{J}_P + \mathrm{J}_M + \varepsilon_0 \frac{\partial \mathrm{E}}{\partial t} = \mathrm{J} + \varepsilon_0 \frac{\partial \mathrm{E}}{\partial t},$$

where current density J_D is the *displacement current*, and J is the current density contribution actually due to movement of charges, both free and bound. Because $\nabla \cdot \mathrm{D} = \rho$, the charge continuity issue with Ampère's original formulation is no longer a problem. Because of the term in $\varepsilon_0 \partial \mathrm{E} / \partial t$, wave propagation in free space now is possible.

With the addition of the displacement current, Maxwell was able to hypothesize (correctly) that light was a form of electromagnetic wave.

Proof of Equivalence

Ampère's law in Cgs Units

In cgs units, the integral form of the equation, including Maxwell's correction, reads

$$\oint_C \mathrm{B} \cdot d\ell = \frac{1}{c} \iint_S \left(4\pi \mathrm{J} + \frac{\partial \mathrm{E}}{\partial t} \right) \cdot d\mathrm{S}$$

where c is the speed of light. The differential form of the equation (again, including Maxwell's correction) is

$$\nabla \times \mathrm{B} = \frac{1}{c} \left(4\pi \mathrm{J} + \frac{\partial \mathrm{E}}{\partial t} \right).$$

Electrical Network

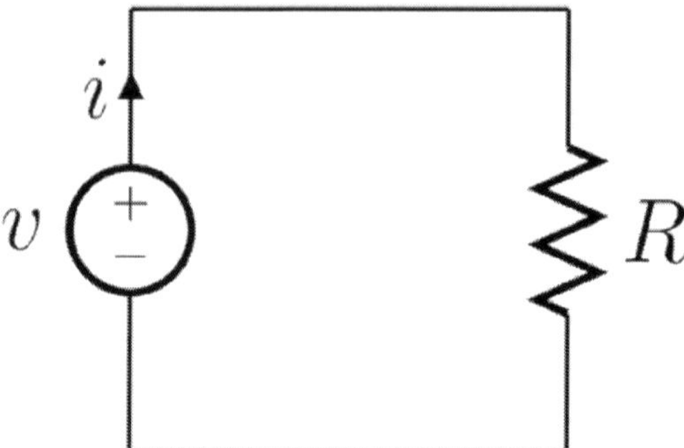

Figure: *A simple electric circuit made up of a voltage source and a resistor. Here,* $V = iR$ *, according to Ohm's Law.*

An electrical network is an interconnection of electrical components (e.g. batteries, resistors, inductors, capacitors, switches) or a model of such an interconnection, consisting of electrical elements (e.g. voltage sources, current sources, resistances, inductances, capacitances). An electrical circuit is a network consisting of a closed loop, giving a return path for the current. Linear electrical networks, a special type consisting only of sources (voltage or current), linear lumped elements (resistors, capacitors, inductors), and linear distributed elements (transmission lines), have the property that signals are linearly superimposable. They are thus more easily analyzed, using powerful frequency domain methods such as Laplace transforms, to determine DC response, AC response, and transient response.

A resistive circuit is a circuit containing only resistors and ideal current and voltage sources. Analysis of resistive circuits is less complicated than analysis of circuits containing capacitors and inductors. If the sources are constant (DC) sources, the result is a DC circuit.

A network that contains active electronic components is known as an *electronic circuit.* Such networks are generally nonlinear and require more complex design and analysis tools.

Classification

By passivity

An active network is a network that consists of at least one active source like a voltage source or current source.

A passive network is a network which does not contain any active device.

By linearity

A network is linear if its signals obey the principle of superposition; otherwise it is non-linear. A linear network will be composed entirely of independent sources, linear dependent sources and linear passive elements.

Classification of Sources

Sources can be classified as independent sources and dependent sources

Independent Sources

Ideal Independent Source maintains same voltage or current regardless of the other elements present in the circuit. Its value is either constant (DC) or sinusoidal (AC). The strength of voltage or current is not changed by any variation in connected network.

Dependent Sources

Dependent Sources depend upon a particular element of the circuit for delivering the power or voltage or current depending upon the type of source it is.

Electrical Laws

A number of electrical laws apply to all electrical networks. These include:

- Kirchhoff's current law: The sum of all currents entering a node is equal to the sum of all currents leaving the node.
- Kirchhoff's voltage law: The directed sum of the electrical potential differences around a loop must be zero.
- Ohm's law: The voltage across a resistor is equal to the product of the resistance and the current flowing through it.
- Norton's theorem: Any network of voltage or current sources and resistors is electrically equivalent to an ideal current source in parallel with a single resistor.
- Thévenin's theorem: Any network of voltage or current sources and resistors is electrically equivalent to a single voltage source in series with a single resistor.
- superposition theorem: In a linear network with several independent sources, the response in a particular branch when all the sources are acting simultaneously is equal to the linear sum of individual responses calculated by talking one independent source at a time.

Design Methods

To design any electrical circuit, either analog or digital, electrical engineers need to be able to predict the voltages and currents at all places within the circuit. Linear circuits, that is, circuits with the same input and output frequency, can be analyzed by hand using complex number theory. Other circuits can only be analyzed with specialized software programs or estimation techniques such as the piecewise-linear model.

Circuit simulation software, such as HSPICE, and languages such as VHDL-AMS and verilog-AMS allow engineers to design circuits without the time, cost and risk of error involved in building circuit prototypes.

Other more complex laws may be needed if the network contains nonlinear or reactive components. Non-linear self-regenerative heterodyning systems can be approximated. Applying these laws results in a set of simultaneous equations that can be solved either algebraically or numerically.

Network Simulation Software

More complex circuits can be analyzed numerically with software such as SPICE or GNUCAP, or symbolically using software such as SapWin.

Linearization Around Operating Point

When faced with a new circuit, the software first tries to find a steady state solution, that is, one where all nodes conform to Kirchhoff's Current Law *and* the voltages across and through each element of the circuit conform to the voltage/current equations governing that element.

Once the steady state solution is found, the operating points of each element in the circuit are known. For a small signal analysis, every non-linear element can be linearized around its operation point to obtain the small-signal estimate of the voltages and currents. This is an application of Ohm's Law. The resulting linear circuit matrix can be solved with Gaussian elimination.

Piecewise-linear Approximation

Software such as the PLECS interface to Simulink uses piecewise-linear approximation of the equations governing the elements of a circuit. The circuit is treated as a completely linear network of ideal diodes. Every time a diode switches from on to off or vice versa, the configuration of the linear network changes. Adding more detail to the approximation of equations increases the accuracy of the simulation, but also increases its running time.

Electrical Element

Electrical elements are conceptual abstractions representing idealized electrical components, such as resistors, capacitors, and inductors, used in the analysis of electrical networks. Any electrical network can be analysed as multiple, interconnected electrical elements in a schematic diagram or circuit diagram, each of which affects the voltage in the network or current through the network.

These ideal electrical elements represent real, physical electrical or electronic components but they do not exist physically and they are assumed to have ideal properties according to a lumped element model, while components are objects with less than ideal properties, a degree of uncertainty in their values and some degree of nonlinearity, each of which may require a combination of multiple electrical elements in order to approximate its function.

Circuit analysis using electric elements is useful for understanding many practical electrical networks using components. By analyzing the way a network is affected by its individual elements it is possible to estimate how a real network will behave.

One-port Elements

Only nine types of element (memristor not included), five passive and four active, are required to model any electrical component or circuit. Each element is defined by a relation between the state variables of the network: current, I; voltage, V, charge, Q; and magnetic flux, Φ.

- Two sources:
  - o Current source, measured in amperes – produces a current in a conductor. Affects charge according to the relation $dQ = -I\,dt$.
  - o Voltage source, measured in volts – produces a potential difference between two points. Affects magnetic flux according to the relation $d\Phi = V\,dt$.

Φ in this relationship does not necessarily represent anything physically meaningful. In the case of the current generator, Q, the time integral of current, represents the quantity of electric charge physically delivered by the generator. Here Φ is the time integral of voltage but whether or not that represents a physical quantity depends on the nature of the voltage source. For a voltage generated by magnetic induction it is meaningful, but for an electrochemical source, or a voltage that is the output of another circuit, no physical meaning is attached to it.

Both these elements are necessarily non-linear elements.

- Three passive elements:
  - Resistance R, measured in ohms – produces a voltage proportional to the current flowing through the element. Relates voltage and current according to the relation $dV = RdI$.
  - Capacitance C, measured in farads – produces a current proportional to the rate of change of voltage across the element. Relates charge and voltage according to the relation $dQ = CdV$.
  - Inductance L, measured in henries – produces the magnetic flux proportional to the rate of change of current through the element. Relates flux and current according to the relation $d\Phi = LdI$.
- Four abstract active elements:
  - Voltage-controlled voltage source (VCVS) Generates a voltage based on another voltage with respect to a specified gain. (has infinite input impedance and zero output impedance).
  - Voltage-controlled current source (VCCS) Generates a current based on a voltage elsewhere in the circuit, with respect to a specified gain, used to model field-effect transistors and vacuum tubes (has infinite input impedance and infinite output impedance).

    The gain is characterised by a transfer conductance which will have units of siemens.
  - Current-controlled voltage source (CCVS) Generates a voltage based on an input current elsewhere in the circuit with respect to a specified gain. (has zero input impedance and zero output impedance). The gain is characterised by a transfer impedance which will have units of ohms.
  - Current-controlled current source (CCCS) Generates a current based on an input current and a specified gain. Used to model bipolar junction transistors. (Has zero input impedance and infinite output impedance).

These four elements are examples of two-port elements.

Non-linear Elements

In reality, all circuit components are non-linear and can only be approximated to linear over a certain range. To more exactly describe the passive elements, their constitutive relation is used instead of simple proportionality. From any two of the circuit variables there are six constitutive relations that can be formed. From this it is supposed that there is a theoretical fourth passive element since there are only five elements in total (not including the various dependent sources) found in linear network analysis.

This additional element is called memristor. It only has any meaning as a time-dependent non-linear element; as a time-independent linear element it reduces to a regular resistor. The constitutive relations of the passive elements are given by;

- Resistance: constitutive relation defined as $f(V,I)=0$.
- Capacitance: constitutive relation defined as $f(V,Q)=0$.
- Inductance: constitutive relation defined as $f(\Phi,I)=0$.
- Memristance: constitutive relation defined as $f(\Phi,Q)=0$.

where $f(x,y)$ is an arbitrary function of two variables.

In some special cases the constitutive relation simplifies to a function of one variable. This is the case for all linear elements, but also for example, an ideal diode, which in circuit theory terms is a non-linear resistor, has a constitutive relation of the form $V=f(I)$. Both independent voltage, and independent current sources can be considered non-linear resistors under this definition.

The fourth passive element, the memristor, was proposed by Leon Chua in a 1971 paper, but a physical component demonstrating memristance was not created until thirty-seven years later. It was reported on April 30, 2008, that a working memristor had been developed by a team at HP Labs led by scientist R. Stanley Williams. With the advent of the memristor, each pairing of the four variables can now be related. Because memristors are time-variant by definition, they are not included in linear time-invariant (LTI) circuit models.

There are also two special non-linear elements which are sometimes used in analysis but which are not the ideal counterpart of any real component:

- Nullator: defined as $V - I - 0$
- Norator: defined as an element which places no restrictions on voltage and current whatsoever.

These are sometimes used in models of components with more than two terminals: transistors for instance.

Two-port Elements

All the above are two-terminal, or one-port, elements with the exception of the dependent sources. There are two lossless, passive, linear two-port elements that are normally introduced into network analysis. Their constitutive relations in matrix notation are;

Transformer

$$\begin{bmatrix} V_1 \\ I_2 \end{bmatrix} = \begin{bmatrix} 0 & n \\ -n & 0 \end{bmatrix} \begin{bmatrix} I_1 \\ V_2 \end{bmatrix}$$

Gyrator

$$\begin{bmatrix} V_1 \\ V_2 \end{bmatrix} = \begin{bmatrix} 0 & -r \\ r & 0 \end{bmatrix} \begin{bmatrix} I_1 \\ I_2 \end{bmatrix}$$

The transformer maps a voltage at one port to a voltage at the other in a ratio of n. The current between the same two port is mapped by $1/n$. The gyrator, on the other hand, maps a voltage at one port to a current at the other. Likewise, currents are mapped to voltages. The quantity r in the matrix is in units of resistance. The gyrator is a necessary element in analysis because it is not reciprocal. Networks built from the basic linear elements only are obliged to be reciprocal and so cannot be used by themselves to represent a non-reciprocal system. It is not essential, however, to have both the transformer and gyrator. Two gyrators in cascade are equivalent to a transformer but the transformer is usually retained for convenience.

Introduction of the gyrator also makes either capacitance or inductance non-essential since a gyrator terminated with one of these at port 2 will be equivalent to the other at port 1. However, transformer, capacitance and inductance are normally retained in analysis because they are the ideal properties of the basic physical components transformer, inductor and capacitor whereas a practical gyrator must be constructed as an active circuit.

Examples

The following are examples of representation of components by way of electrical elements.

- On a first degree of approximation, a battery is represented by a voltage source. A more refined model also includes a resistance in series with the voltage source, to represent the battery's internal resistance (which results in the battery heating and the voltage dropping when in use). A current source in parallel may be added to represent its leakage (which discharges the battery over a long period of time).

- On a first degree of approximation, a resistor is represented by a resistance. A more refined model also includes a series inductance, to represent the effects of its lead inductance (resistors constructed as a spiral have more significant inductance). A capacitance in parallel may be added to represent the capacitive effect of the proximity of the resistor leads to each other. A wire can be represented as a low-value resistor
- Current sources are more often used when representing semiconductors. For example, on a first degree of approximation, a bipolar transistor may be represented by a variable current source that is controlled by the input current.

Chapter 3

Electronic Component

An electronic component is any basic discrete device or physical entity in an electronic system used to affect electrons or their associated fields. Electronic components are mostly industrial products, available in a singular form and are not to be confused with electrical elements, which are conceptual abstractions representing idealized electronic components.

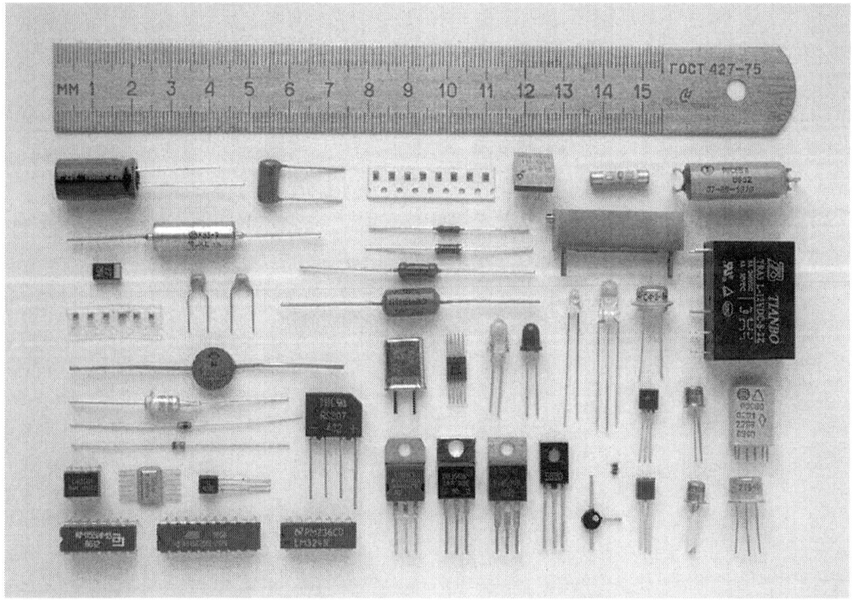

Figure: *Various electronic components*

Electronic components have two or more electrical terminals (or *leads*) aside from antennas which may only have one terminal. These leads connect, usually soldered to a printed circuit board, to create an electronic circuit (a discrete circuit) with a particular function (for example an amplifier, radio receiver, or oscillator). Basic electronic components may be packaged discretely, as arrays or networks of like components, or integrated inside of packages such as semiconductor integrated circuits, hybrid integrated circuits, or thick film devices. The following list of electronic components focuses on the discrete version of these components, treating such packages as components in their own right.

Classification

A component may be classified as passive, active, or electromechanic. The strict physics definition treats passive components as ones that cannot supply energy themselves, whereas a battery would be seen as an active component since it truly acts as a source of energy.

However, electronic engineers who perform circuit analysis use a more restrictive definition of passivity. When only concerned with the energy of signals, it is convenient to ignore the so-called DC circuit and pretend that the power supplying components such as transistors or integrated circuits is absent (as if each such component had its own battery built in), though it may in reality be supplied by the DC circuit. Then, the analysis only concerns the AC circuit, an abstraction that ignores DC voltages and currents (and the power associated with them) present in the real-life circuit. This fiction, for instance, lets us view an oscillator as "producing energy" even though in reality the oscillator consumes even more energy from a DC power supply, which we have chosen to ignore. Under that restriction, we define the terms as used in circuit analysis as:

- Active components rely on a source of energy (usually from the DC circuit, which we have chosen to ignore) and usually can inject power into a circuit, though this is not part of the definition. Active components include amplifying components such as transistors, triode vacuum tubes (valves), and tunnel diodes.
- Passive components can't introduce net energy into the circuit. They also can't rely on a source of power, except for what is available from the (AC) circuit they are connected to. As a consequence they can't amplify (increase the power of a signal), although they may increase a voltage or current (such as is done by a transformer or resonant circuit). Passive components include two-terminal components such as resistors, capacitors, inductors, and transformers.

- Electromechanical components can carry out electrical operations by using moving parts or by using electrical connections

Most passive components with more than two terminals can be described in terms of two-port parameters that satisfy the principle of reciprocity—though there are rare exceptions. In contrast, active components (with more than two terminals) generally lack that property.

Active Components

Semiconductors

Diodes: Conduct electricity easily in one direction, among more specific behaviours.

- Diode, Rectifier, Bridge rectifier
- Schottky diode, hot carrier diode – super fast diode with lower forward voltage drop
- Zener diode – Passes current in reverse direction to provide a constant voltage reference
- Transient voltage suppression diode (TVS), Unipolar or Bipolar – used to absorb high-voltage spikes
- Varactor, Tuning diode, Varicap, Variable capacitance diode – A diode whose AC capacitance varies according to the DC voltage applied.
- Light-emitting diode (LED) – A diode that emits light
- LASER diode – A semiconductor laser
- Photodiode – Passes current in proportion to incident light
  - o Avalanche Photodiode with internal gain
  - o Solar Cell, photovoltaic cell, PV array or panel, produces power from light
- DIAC (Diode for Alternating Current), Trigger Diode, SIDAC) – Often used to trigger an SCR
- Constant-current diode
- Peltier cooler – A semiconductor heat pump

Transistors: Transistors were considered the invention of the twentieth century that changed electronic circuits forever. A transistor is a semiconductor device used to amplify and switch electronic signals and electrical power.

- Transistors
    - o Bipolar junction transistor (BJT, or simply "transistor") – NPN or PNP
        - – Photo transistor – Amplified photodetector
    - o Darlington transistor – NPN or PNP
        - – Photo Darlington – Amplified photodetector
    - o Sziklai pair (Compound transistor, complementary Darlington)
- Field-effect transistor (FET)
    - o JFET (Junction Field-Effect Transistor) – N-CHANNEL or P-CHANNEL
    - o MOSFET (Metal Oxide Semiconductor FET) – N-CHANNEL or P-CHANNEL
    - o MESFET (MEtal Semiconductor FET)
    - o HEMT (High electron mobility transistor)
- Thyristors
    - o Silicon-controlled rectifier (SCR) – Passes current only after triggered by a sufficient control voltage on its gate
    - o TRIAC (TRIode for Alternating Current) – Bidirectional SCR
    - o Unijunction transistor (UJT)
    - o Programmable Unijunction transistor (PUT)
    - o SIT (Static induction transistor)
    - o SITh (Static induction thyristor)
- Composite transistors
    - o IGBT (Insulated-gate bipolar transistor)

Integrated Circuits:

- Digital
- Analog
    - o Hall effect sensor –senses a magnetic field
    - o Current sensor – Senses a current through it

Optoelectronic Devices

- Optoelectronics
    - o Opto-Isolator, Opto-Coupler, Photo-Coupler – Photodiode, BJT, JFET, SCR, TRIAC, Zero-crossing TRIAC, Open collector IC, CMOS IC, Solid state relay (SSR)

- o Opto switch, Opto interrupter, Optical switch, Optical interrupter, Photo switch, Photo interrupter
- o LED display – Seven-segment display, Sixteen-segment display, Dot-matrix display

Display Technologies

Current:

- Filament lamp (indicator lamp)
- Vacuum fluorescent display (VFD) (preformed characters, 7 segment, starburst)
- Cathode ray tube (CRT) (dot matrix scan, radial scan (e.g. radar), arbitrary scan (e.g. oscilloscope)) (monochrome & colour)
- LCD (preformed characters, dot matrix) (passive, TFT) (monochrome, colour)
- Neon (individual, 7 segment display)
- LED (individual, 7 segment display, starburst display, dot matrix)
- Flap indicator (numeric, preprinted messages)
- Plasma display (dot matrix)

Obsolete:

- Filament lamp 7 segment display (aka 'minitron')
- Nixie Tube
- Dekatron (aka glow transfer tube)
- Magic eye tube indicator
- Penetron (a 2 colour see-through CRT)

Vacuum Tubes (Valves)

A vacuum tube is based on current conduction through a vacuum.

- Diode or rectifier tube

Amplifying tubes

- Triode
- Tetrode
- Pentode
- Hexode
- Pentagrid
- Octode
- Microwave tubes

- o Klystron
- o Magnetron
- o Traveling-wave tube

Optical detectors or emitters

- Phototube or Photodiode – tube equivalent of semiconductor photodiode
- Photomultiplier tube – Phototube with internal gain
- Cathode ray tube (CRT) or television picture tube
- Vacuum fluorescent display (VFD) – Modern non-raster sort of small CRT display
- Magic eye tube – Small CRT display used as a tuning metre (obsolete)
- X-ray tube – Produces x-rays

Discharge Devices

- Gas discharge tube

Obsolete:

- Mercury arc rectifier
- Voltage regulator tube
- Nixie tube
- Thyratron
- Ignitron

Power Sources

Sources of electrical power:

- Battery – acid- or alkali-based power supply
- Fuel cell – an electrochemical generator
- Power supply – usually a mains hook-up
- Photo voltaic device – generates electricity from light
- Thermo electric generator – generates electricity from temperature gradients
- Electrical generator – an electromechanical power source
- Piezoelectric pressure - creates electricity from mechanical strain
- Physically carrying electrons - Van de Graaff generator or essentially creating voltage from friction

Passive Components

Resistors

Pass current in proportion to voltage (Ohm's law) and oppose current.

- Resistor – fixed value
  - Power resistor – larger to safely dissipate heat generated
  - SIP or DIP resistor network – array of resistors in one package
- Variable resistor
  - Rheostat – two-terminal variable resistor (often for high power)
  - Potentiometre – three-terminal variable resistor (variable voltage divider)
  - Trim pot – Small potentiometre, usually for internal adjustments
  - Thermistor – thermally sensitive resistor whose prime function is to exhibit a large, predictable and precise change in electrical resistance when subjected to a corresponding change in body temperature.
  - Humistor – humidity-varied resistor
  - Photoresistor
  - Memristor
  - Varistor, Voltage Dependent Resistor, MOV – Passes current when excessive voltage is present
- Resistance wire, Nichrome wire – wire of high-resistance material, often used as a heating element
- Heater – heating element

Capacitors

Capacitors store and release electrical charge. They are used for filtering power supply lines, tuning resonant circuits, and for blocking DC voltages while passing AC signals, among numerous other uses.

- Capacitor
  - Integrated capacitors
    - MIS capacitor
    - Trench capacitor
  - Fixed capacitors

  - Ceramic capacitor
  - Film capacitor
  - Electrolytic capacitor
  - Aluminum electrolytic capacitor
  - Tantalum electrolytic capacitor
  - Niobium electrolytic capacitor
  - Polymer capacitor
  - OS-CON
  - Electric double-layer capacitor
  - Nanoionic supercapacitor
  - Lithium-ion capacitor
  - Mica capacitor
  - Vacuum capacitor
- o Variable capacitor – adjustable capacitance
  - Tuning capacitor – variable capacitor for tuning a radio, oscillator, or tuned circuit
  - Trim capacitor– small variable capacitor is usually for slight internal adjustments made with a small screw driver turned into the right position.
  - Vacuum variable capacitor
- o Capacitors for special applications
  - Power capacitor
  - Safety capacitor
  - Filter capacitor
  - Light-emitting capacitor
  - Motor capacitor
  - Photoflash capacitor
  - Reservoir capacitor
- o Capacitor network (array)

- Varicap diode – AC capacitance varies according to the DC voltage applied

Magnetic (Inductive) Devices

Electrical components that use magnetism in the storage and release of electrical charge through current:

- Inductor, coil, choke

- Variable inductor
- Saturable Inductor
- Transformer
- Magnetic amplifier (toroid)
- ferrite impedances, beads
- Motor / Generator
- Solenoid
- Loudspeaker and microphone

Memristor

Electrical components that pass charge in proportion to magnetism or magnetic flux, and have the ability to retain a previous resistive state, hence the name of Memory plus Resistor.

- Memristor

Networks

Components that use more than one type of passive component:

- RC network – forms an RC circuit, used in snubbers
- LC Network – forms an LC circuit, used in tunable transformers and RFI filters.

Transducers, Sensors, Detectors

1. Transducers generate physical effects when driven by an electrical signal, or vice-versa.
2. Sensors (detectors) are transducers that react to environmental conditions by changing their electrical properties or generating an electrical signal.
3. The transducers listed here are single electronic components (as opposed to complete assemblies), and are passive.

   Only the most common ones are listed here.

- Audio
  - o Loudspeaker – Magnetic or piezoelectric device to generate full audio
  - o Buzzer – Magnetic or piezoelectric sounder to generate tones
- Position, motion
  - o Linear variable differential transformer (LVDT) – Magnetic – detects linear position

   - Rotary encoder, Shaft Encoder – Optical, magnetic, resistive or switches – detects absolute or relative angle or rotational speed
   - Inclinometre – Capacitive – detects angle with respect to gravity
   - Motion sensor, Vibration sensor
   - Flow metre – detects flow in liquid or gas
- Force, torque
   - Strain gauge – Piezoelectric or resistive – detects squeezing, stretching, twisting
   - Accelerometre – Piezoelectric – detects acceleration, gravity
- Thermal
   - Thermocouple, thermopile – Wires that generate a voltage proportional to delta temperature
   - Thermistor – Resistor whose resistance changes with temperature, up PTC or down NTC
   - Resistance Temperature Detector (RTD) – Wire whose resistance changes with temperature
   - Bolometre – Device for measuring the power of incident electromagnetic radiation
   - Thermal cutoff – Switch that is opened or closed when a set temperature is exceeded
- Magnetic field
   - Magnetometre, Gauss metre
- Humidity
   - Hygrometre
- Electromagnetic, light
   - Photo resistor – Light dependent resistor (LDR)

Antennas

Antennas transmit or receive radio waves

- Elemental dipole
- Yagi
- Phased array
- Loop antenna
- Parabolic dish

- Log-periodic dipole array
- Biconical
- Feedhorn

Assemblies, Modules

Multiple electronic components assembled in a device that is in itself used as a component

- Oscillator
- Display devices
  - o Liquid crystal display (LCD)
  - o Digital voltmetres
- Filter

Prototyping Aids

- Wire-wrap
- Breadboard

Electromechanical

Piezoelectric Devices, Crystals, Resonators

Passive components that use piezoelectric effect:

- Components that use the effect to generate or filter high frequencies
  - o Crystal – a ceramic crystal used to generate precise frequencies
  - o Ceramic resonator – Is a ceramic crystal used to generate semi-precise frequencies
  - o Ceramic filter – Is a ceramic crystal used to filter a band of frequencies such as in radio receivers
  - o surface acoustic wave (SAW) filters
- Components that use the effect as mechanical transducers.
  - o Ultrasonic motor – Electric motor that uses the piezoelectric effects
  - o For piezo buzzers and microphones

Terminals and Connectors

Devices to make electrical connection

- Terminal
- Connector

- o Socket
- o Screw terminal, Terminal Blocks
- o Pin header

Cable Assemblies

Cables with connectors or terminals at their ends

- Power cord
- Patch cord
- Test lead

Figure: *2 different tactile switches*

Switches

Standard Symbols: On a circuit diagram, electronic devices are represented by conventional symbols. Reference designators are applied to the symbols to identify the component.

Series and Parallel Circuits

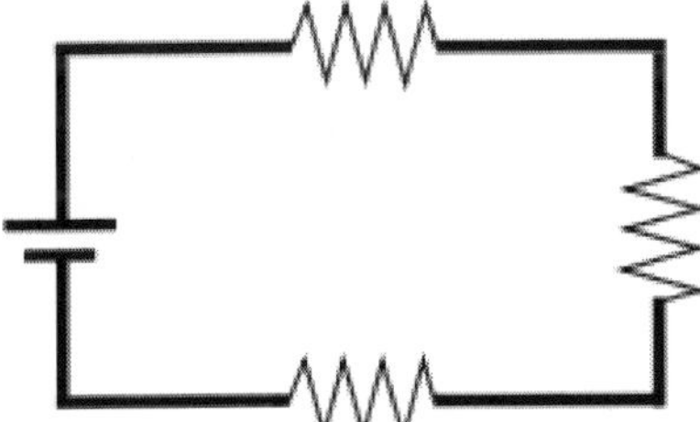

Figure: *A series circuit with a voltage source (such as a battery) and 3 resistors*

Components of an electrical circuit or electronic circuit can be connected in many different ways. The two simplest of these are called series and parallel and occur very frequently. Components connected in series are connected along a single path, so the same current flows through all of the components. Components connected in parallel are connected so the same voltage is applied to each component. A circuit composed solely of components connected in series is known as a series circuit; likewise, one connected completely in parallel is known as a parallel circuit.

In a series circuit, the current through each of the components is the same, and the voltage across the circuit is the sum of the voltages across each component.

In a parallel circuit, the voltage across each of the components is the same, and the total current is the sum of the currents through each component.

As an example, consider a very simple circuit consisting of four light bulbs and one 6 V battery. If a wire joins the battery to one bulb, to the next bulb, to the next bulb, to the next bulb, then back to the battery, in one continuous loop, the bulbs are said to be in series. If each bulb is wired to the battery in a separate loop, the bulbs are said to be in parallel.

If the four light bulbs are connected in series, there is same current through all of them, and the voltage drop is 1.5 V across each bulb, which may not be sufficient to make them glow. If the light bulbs are connected in parallel, the currents through the light bulbs combine to form the current in the battery, while the voltage drop is 6.0 V across each bulb and they all glow.

In a series circuit, every device must function for the circuit to be complete. One bulb burning out in a series circuit breaks the circuit. In parallel circuits, each light has its own circuit, so all but one light could be burned out, and the last one will still function.

Series circuits

Series circuits are sometimes called *current*-coupled or daisy chain-coupled. The current in a series circuit goes through every component in the circuit. Therefore, all of the components in a series connection carry the same current. There is only one path in a series circuit in which the current can flow.

A series circuit's main disadvantage or advantage, depending on its intended role in a product's overall design, is that because there is only one path in which its current can flow, opening or breaking a series circuit at any point causes the entire circuit to "open" or stop operating. For example, if even one of the light bulbs in an older-style string of Christmas tree lights burns out or is removed, the entire string becomes inoperable until the bulb is replaced.

Current

$$I = I_1 = I_2 = \ldots = I_n$$

In a series circuit the current is the same for all elements.

Resistors

The total resistance of resistors in series is equal to the sum of their individual resistances:

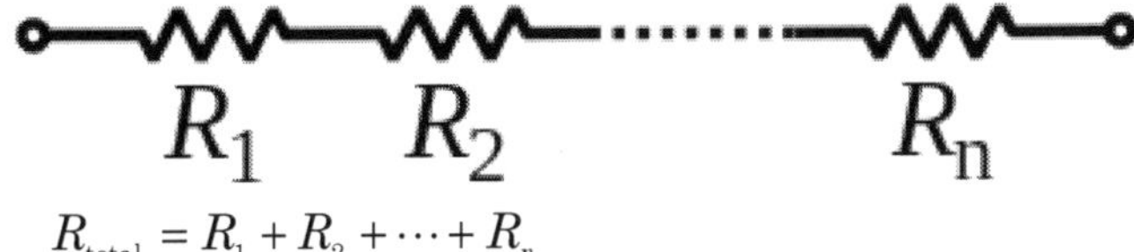

$$R_{\text{total}} = R_1 + R_2 + \cdots + R_n$$

Electrical conductance presents a reciprocal quantity to resistance. Total conductance of a series circuits of pure resistors, therefore, can be calculated from the following expression:

$$\frac{1}{G_{\text{total}}} = \frac{1}{G_1} + \frac{1}{G_2} + \cdots + \frac{1}{G_n}.$$

For a special case of two resistors in series, the total conductance is equal to:

$$G_{total} = \frac{G_1 G_2}{G_1 + G_2}.$$

Inductors

Inductors follow the same law, in that the total inductance of non-coupled inductors in series is equal to the sum of their individual inductances:

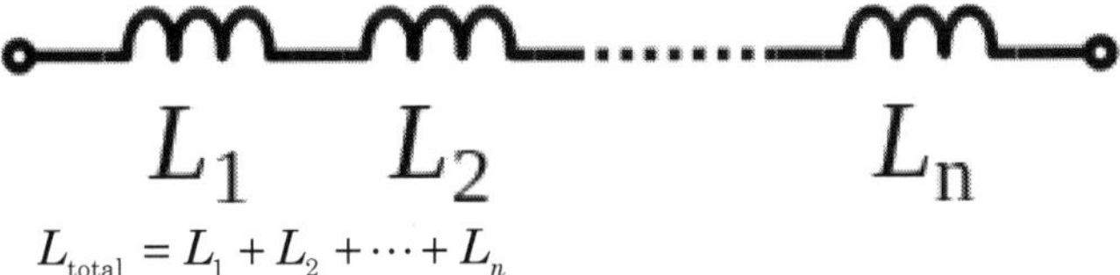

$$L_{\text{total}} = L_1 + L_2 + \cdots + L_n$$

However, in some situations it is difficult to prevent adjacent inductors from influencing each other, as the magnetic field of one device couples with the windings of its neighbours. This influence is defined by the mutual inductance M. For example, if two inductors are in series, there are two possible equivalent inductances depending on how the magnetic fields of both inductors influence each other.

When there are more than two inductors, the mutual inductance between each of them and the way the coils influence each other complicates the calculation. For a larger number of coils the total combined inductance is given by the sum of all mutual inductances between the various coils including the mutual inductance of each given coil with itself, which we term self-inductance or simply inductance.

For three coils, there are six mutual inductances M_{12}, M_{13}, M_{23} and M_{21}, M_{31} and M_{32}. There are also the three self-inductances of the three coils: M_{11}, M_{22} and M_{33}.

Therefore

$$L_{total} = (M_{11} + M_{22} + M_{33}) + (M_{12} + M_{13} + M_{23}) + (M_{21} + M_{31} + M_{32})$$

By reciprocity $M_{ij} = M_{ji}$ so that the last two groups can be combined. The first three terms represent the sum of the self-inductances of the various coils.

The formula is easily extended to any number of series coils with mutual coupling.

The method can be used to find the self-inductance of large coils of wire of any cross-sectional shape by computing the sum of the mutual inductance of each turn of wire in the coil with every other turn since in such a coil all turns are in series.

Capacitors

Capacitors follow the same law using the reciprocals. The total capacitance of capacitors in series is equal to the reciprocal of the sum of the reciprocals of their individual capacitances:

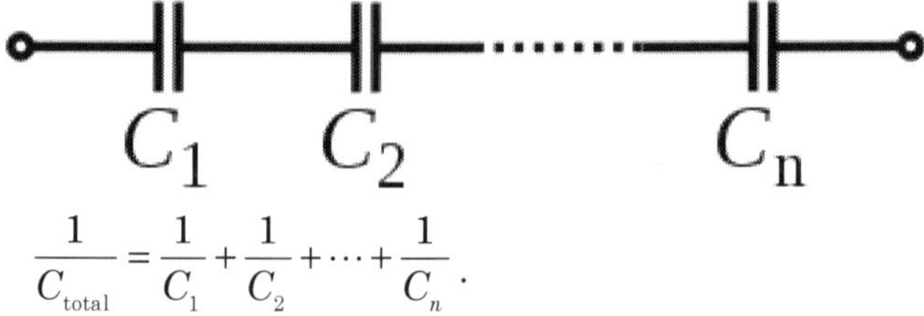

$$\frac{1}{C_{total}} = \frac{1}{C_1} + \frac{1}{C_2} + \cdots + \frac{1}{C_n}.$$

Switches: Two or more switches in series form a logical AND; the circuit only carries current if all switches are 'on'.

Cells and Batteries

A battery is a collection of electrochemical cells. If the cells are connected in series, the voltage of the battery will be the sum of the cell voltages.

For example, a 12 volt car battery contains six 2-volt cells connected in series. Some vehicles, such as trucks, have two 12 volt batteries in series to feed the 24 volt system.

Parallel Circuits

If two or more components are connected in parallel they have the same potential difference (voltage) across their ends. The potential differences across the components are the same in magnitude, and they also have identical polarities. The same voltage is applicable to all circuit components connected in parallel. The total current is the sum of the currents through the individual components, in accordance with Kirchhoff's current law.

Voltage

In a parallel circuit the voltage is the same for all elements.

$$V = V_1 = V_2 = \ldots = V_n$$

Resistors

The current in each individual resistor is found by Ohm's law. Factoring out the voltage gives

$$I_{\text{total}} = V\left(\frac{1}{R_1} + \frac{1}{R_2} + \cdots + \frac{1}{R_n}\right).$$

To find the total resistance of all components, add the reciprocals of the resistances R_i of each component and take the reciprocal of the sum. Total resistance will always be less than the value of the smallest resistance:

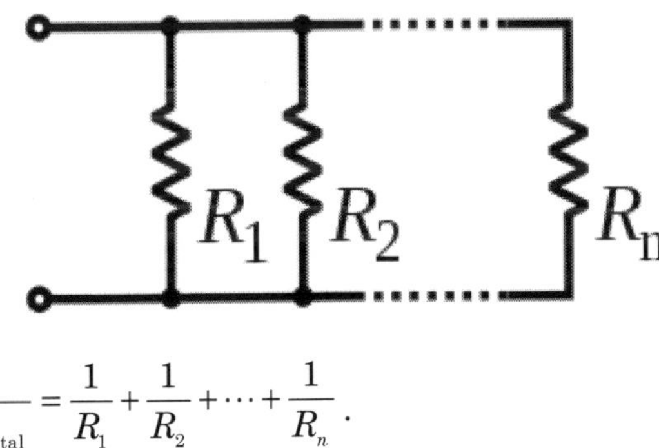

$$\frac{1}{R_{\text{total}}} = \frac{1}{R_1} + \frac{1}{R_2} + \cdots + \frac{1}{R_n}.$$

For only two resistors, the unreciprocated expression is reasonably simple:

$$R_{\text{total}} = \frac{R_1 R_2}{R_1 + R_2}.$$

This sometimes goes by the mnemonic "product over sum".

For N equal resistors in parallel, the reciprocal sum expression simplifies to:

$$\frac{1}{R_{\text{total}}} = \frac{1}{R} \times N\,.$$

and therefore to:

$$R_{\text{total}} = \frac{R}{N}\,.$$

To find the current in a component with resistance R_i, use Ohm's law again:

$$I_i = \frac{V}{R_i}\,.$$

The components divide the current according to their reciprocal resistances, so, in the case of two resistors,

$$\frac{I_1}{I_2} = \frac{R_2}{R_1}\,.$$

An old term for devices connected in parallel is *multiple*, such as a multiple connection for arc lamps.

Since electrical conductance G is reciprocal to resistance, the expression for total conductance of a parallel circuit of resistors reads:

$$G_{\text{total}} = G_1 + G_2 + \cdots + G_n\,.$$

The relations for total conductance and resistance stand in a complementary relationship: the expression for a series connection of resistances is the same as for parallel connection of conductances, and vice versa.

Inductors

Inductors follow the same law, in that the total inductance of non-coupled inductors in parallel is equal to the reciprocal of the sum of the reciprocals of their individual inductances:

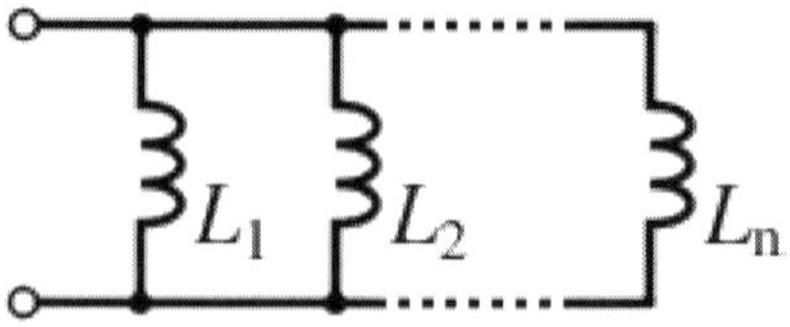

L_1 L_2 L_n

$$\frac{1}{L_{\text{total}}} = \frac{1}{L_1} + \frac{1}{L_2} + \cdots + \frac{1}{L_n}\,.$$

If the inductors are situated in each other's magnetic fields, this approach is invalid due to mutual inductance. If the mutual inductance between two coils in parallel is M, the equivalent inductor is:

$$\frac{1}{L_{\text{total}}} = \frac{L_1 + L_2 - 2M}{L_1 L_2 - M^2}$$

If $L_1 = L_2$

$$L_{total} = \frac{L + M}{2}$$

The sign *M* of depends on how the magnetic fields influence each other. For two equal tightly coupled coils the total inductance is close to that of each single coil. If the polarity of one coil is reversed so that M is negative, then the parallel inductance is nearly zero or the combination is almost non-inductive.

It is assumed in the "tightly coupled" case M is very nearly equal to L. However, if the inductances are not equal and the coils are tightly coupled there can be near short circuit conditions and high circulating currents for both positive and negative values of M, which can cause problems.

More than three inductors becomes more complex and the mutual inductance of each inductor on each other inductor and their influence on each other must be considered.

For three coils, there are three mutual inductances M_{12}, M_{13} and M_{23}. This is best handled by matrix methods and summing the terms of the inverse of the matrix (3 by 3 in this case).

The pertinent equations are of the form: $v_i = \sum_j L_{i,j} \frac{di_j}{dt}$

Capacitors

The total capacitance of capacitors in parallel is equal to the sum of their individual capacitances:

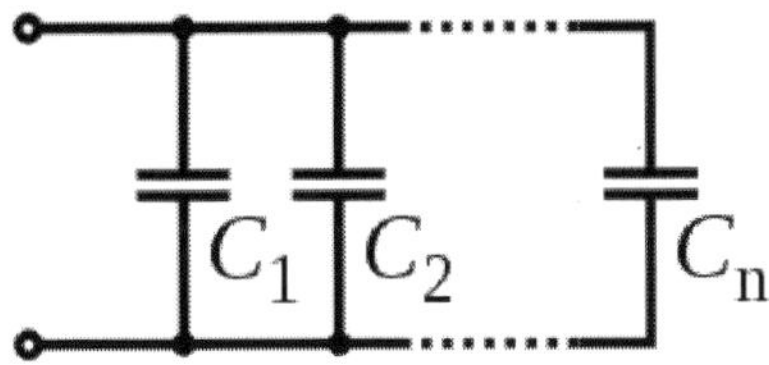

$$C_{\text{total}} = C_1 + C_2 + \cdots + C_n.$$

The working voltage of a parallel combination of capacitors is always limited by the smallest working voltage of an individual capacitor.

Switches

Two or more switches in parallel form a logical OR; the circuit carries current if at least one switch is 'on'.

Cells and Batteries

If the cells of a battery are connected in parallel, the battery voltage will be the same as the cell voltage but the current supplied by each cell will be a fraction of the total current.

For example, if a battery contains four cells connected in parallel and delivers a current of 1 ampere, the current supplied by each cell will be 0.25 ampere. P

arallel-connected batteries were widely used to power the valve filaments in portable radios but they are now rare. Some solar electric systems have batteries in parallel to increase the storage capacity; a close approximation of total amp-hours is the sum of all batteries in parallel.

Combining Conductances

From Kirchhoff's circuit laws we can deduce the rules for combining conductances.

For two conductances G_1 and G_2 in parallel the voltage across them is the same and from Kirchhoff's Current Law the total current is

$$I_{Eq} = I_1 + I_2.$$

Substituting Ohm's law for conductances gives

$$G_{Eq}V = G_1V + G_2V$$

and the equivalent conductance will be,

$$G_{Eq} = G_1 + G_2.$$

For two conductances G_1 and G_2 in series the current through them will be the same and Kirchhoff's Voltage Law tells us that the voltage across them is the sum of the voltages across each conductance, that is,

$$V_{Eq} = V_1 + V_2.$$

Substituting Ohm's law for conductance then gives,

$$\frac{I}{G_{Eq}} = \frac{I}{G_1} + \frac{I}{G_2}$$

which in turn gives the formula for the equivalent conductance,

$$\frac{1}{G_{Eq}} = \frac{1}{G_1} + \frac{1}{G_2}.$$

This equation can be rearranged slightly, though this is a special case that will only rearrange like this for two components.

$$G_{Eq} = \frac{G_1 G_2}{G_1 + G_2}.$$

Notation

The value of two components in parallel is often represented in equations by two vertical lines "| |", borrowing the parallel lines notation from geometry.

$$R_{eq} = R_1 \parallel R_2 = \frac{R_1 R_2}{R_1 + R_2}$$

This simplifies expressions that would otherwise become complicated by expansion of the terms. For instance, the expression $R_1 \parallel R_2 \parallel R_3$ refers to 3 resistors in parallel, while the expanded expression is $\frac{R_1 R_2 R_3}{R_1 R_2 + R_1 R_3 + R_2 R_3}$.

Applications

Most common application of series circuit in consumer electronics is the 9 volt block battery, the fire alarm battery, which is internally built of six batteries, 1.5 volts each.

Series circuits were formerly used for lighting in electric multiple unit trains. For example, if the supply voltage was 600 volts there might be eight 70-volt bulbs in series (total 560 volts) plus a resistor to drop the remaining 40 volts. Series circuits for train lighting were superseded, first by motor-generators, then by solid state devices.

Series resistance can also be applied to the arrangement of blood vessels within a given organ. Each organ is supplied by a large artery, smaller arteries, arterioles, capillaries, and veins arranged in series. The total resistance is the sum of the individual resistances, as expressed by the following equation: $R_{total} = R_{artery} + R_{arterioles} + R_{capillaries}$. The largest proportion of resistance in this series is contributed by the arterioles.

Parallel resistance is illustrated by the circulatory system. Each organ is supplied by an artery that branches off the aorta. The total resistance of this parallel arrangement is expressed by the following equation: $1/R_{total} = 1/R_a + 1/R_b + ... 1/R_n$. R_a, R_b, and R_n are the resistances of the renal, hepatic, and other arteries respectively. The total resistance is less than the resistance of any of the individual arteries.

Circuit Diagram

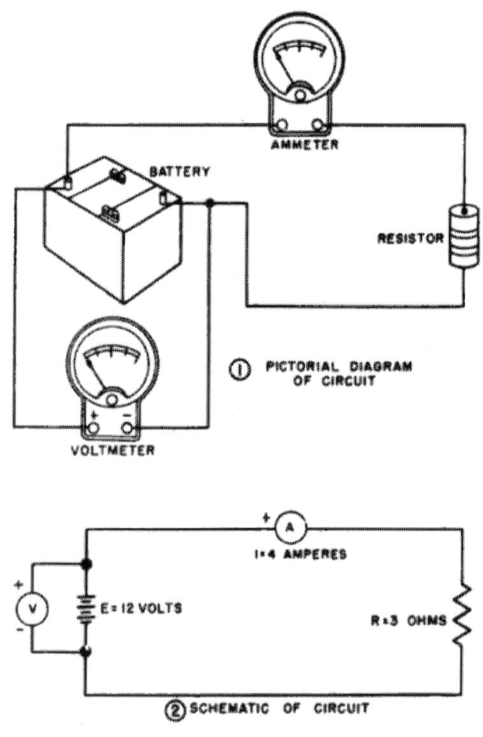

Figure: *Comparison of pictorial and schematic styles of circuit diagrams*

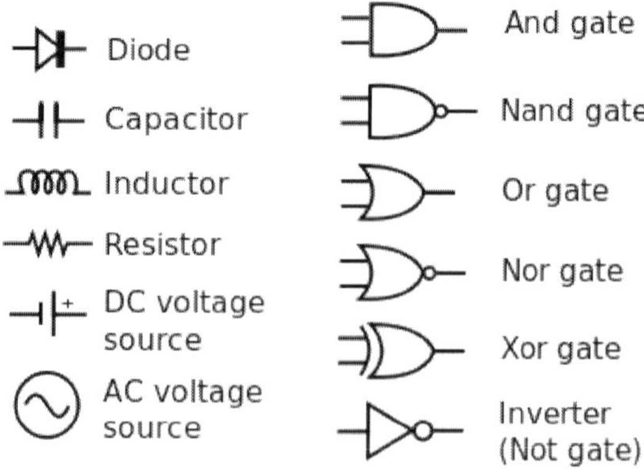

Figure: *Common schematic diagram symbols (US symbols)*

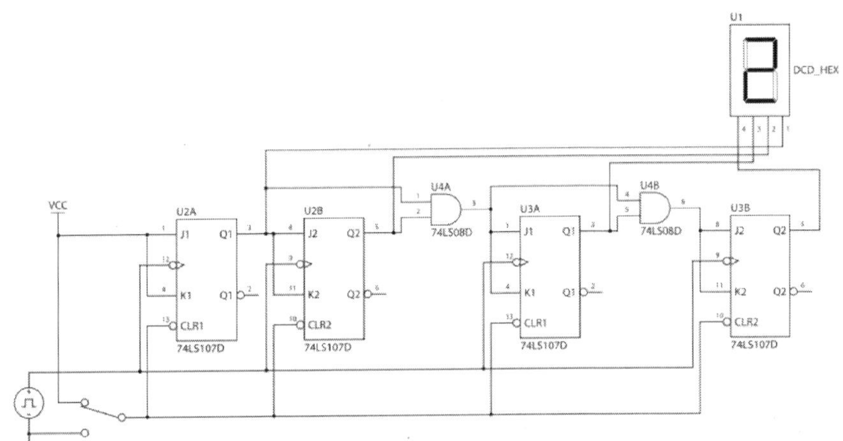

***Figure:** The circuit diagram for a four-bit TTL counter, a type of state machine*

A circuit diagram (also known as an electrical diagram, elementary diagram, or electronic schematic) is a simplified conventional graphical representation of an electrical circuit. A pictorial circuit diagram uses simple images of components, while a schematic diagram shows the components of the circuit as simplified standard symbols; both types show the connections between the devices, including power and signal connections. Arrangement of the components interconnections on the diagram does not correspond to their physical locations in the finished device.

Unlike a block diagram or layout diagram, a circuit diagram shows the actual wire connections being used. The diagram does not show the physical arrangement of components. A drawing meant to depict what the physical arrangement of the wires and the components they connect is called "artwork" or "layout" or the "physical design."

Circuit diagrams are used for the design (circuit design), construction (such as PCB layout), and maintenance of electrical and electronic equipment.

In computer science, circuit diagrams are especially useful when visualizing different expressions using Boolean algebra.

Symbols

Circuit diagrams are pictures with symbols that have differed from country to country and have changed over time, but are now to a large extent internationally standardized. Simple components often had symbols intended to represent some feature of the physical construction of the device. For example, the symbol for a resistor shown here dates back to the days when that component was made from a long piece of wire wrapped in such a manner as to not produce inductance, which would have made it a coil. These wirewound resistors are now used only in high-power applications, smaller resistors being cast from carbon composition (a mixture of carbon and filler) or fabricated as an insulating tube or chip coated with a metal film. The internationally standardized symbol for a resistor is therefore now simplified to an oblong, sometimes with the value in ohms written inside, instead of the zig-zag symbol. A less common symbol is simply a series of peaks on one side of the line representing the conductor, rather than back-and-forth as shown here.

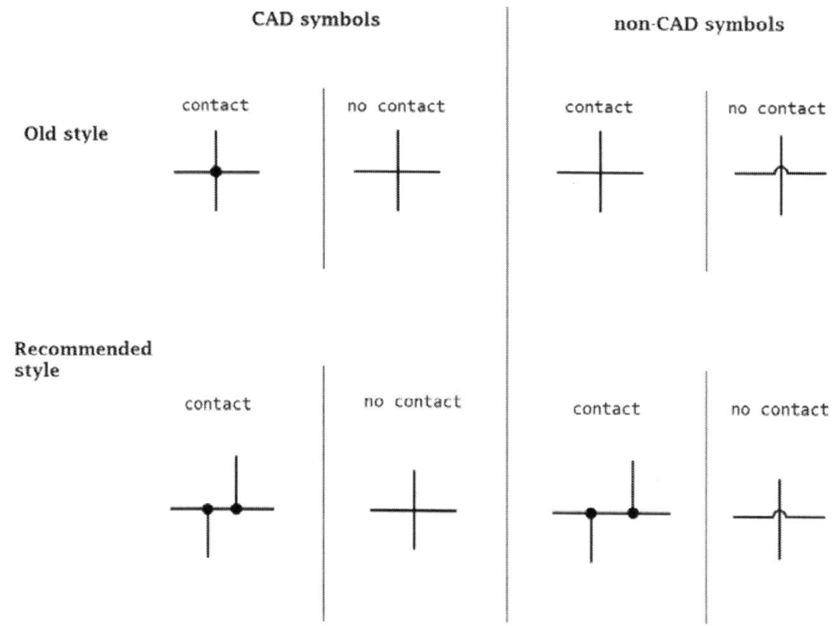

Wire Crossover Symbols for Circuit Diagrams: *Note that the CAD symbol for insulated crossing wires is exactly the same as the older, non-CAD symbol for non-insulated crossing wires. To avoid confusion, the wire "jump" (semi-circle) symbol for insulated wires in non-CAD schematics is recommended (as opposed to using the CAD-style symbol for no connection), so as to avoid confusion with the original, older style symbol, which means the exact opposite. The newer, recommended style for 4-way wire connections in both CAD and non-CAD schematics is to stagger the joining wires into T-junctions.*

The linkages between leads were once simple crossings of lines. With the arrival of computerized drafting, the connection of two intersecting wires was shown by a crossing of wires with a "dot" or "blob" to indicate a connection. At the same time, the crossover was simplified to be the same crossing, but without a "dot". However, there was a danger of confusing the wires that were connected and not connected in this manner, if the dot was drawn too small or accidentally omitted (e.g. the "dot" could disappear after several passes through a copy machine). As such, the modern practice for representing a 4-way wire connection is to draw a straight wire and then to draw the other wires staggered along it with "dots" as connections (see diagram), so as to form two separate T-junctions that brook no confusion and are clearly not a crossover.

For crossing wires that are insulated from one another, a small semi-circle symbol is commonly used to show one wire "jumping over" the other wire (similar to how jumper wires are used).

A common, hybrid style of drawing combines the T-junction crossovers with "dot" connections and the wire "jump" semi-circle symbols for insulated crossings. In this manner, a "dot" that is too small to see or that has accidentally disappeared can still be clearly differentiated from a "jump".

On a circuit diagram, the symbols for components are labelled with a descriptor or reference designator matching that on the list of parts. For example, C1 is the first capacitor, L1 is the first inductor, Q1 is the first transistor, and R1 is the first resistor (note that this is not written as a subscript, as in R1, L1,...). Often the value or type designation of the component is given on the diagram beside the part, but detailed specifications would go on the parts list.

Detailed rules for reference designations are provided in the International standard IEC 61346.

Organization of Drawings

It is a usual although not universal convention that schematic drawings are organized on the page from left to right and top to bottom in the same sequence as the flow of the main signal or power path. For example, a schematic for a radio receiver might start with the antenna input at the left of the page and end with the loudspeaker at the right. Positive power supply connections for each stage would be shown towards the top of the page, with grounds, negative supplies, or other return paths towards the bottom. Schematic drawings intended for maintenance may have the principal signal paths highlighted to assist in understanding the signal flow through the circuit. More complex devices have multi-page schematics and must rely on cross-reference symbols to show the flow of signals between the different sheets of the drawing.

Detailed rules for the preparation of circuit diagrams (and other document kinds used in electrotechnology) are provided in the International standard IEC 61082-1.

Relay logic line diagrams (also called ladder logic diagrams) use another common standardized convention for organizing schematic drawings, with a vertical power supply "rail" on the left and another on the right, and components strung between them like the rungs of a ladder.

Artwork

Once the schematic has been made, it is converted into a layout that can be fabricated onto a printed circuit board (PCB). The layout is usually started by the process of schematic capture. The result is what is known as a rat's nest. The rat's nest is a jumble of wires (lines) criss-crossing each other to their destination nodes. These wires are routed either manually or by the use of electronics design automation (EDA) tools. The EDA tools arrange and rearrange the placement of components and find paths for tracks to connect various nodes. This results in the final layout artwork for the integrated circuit or printed circuit board.

A generalized design flow would be as follows:

Schematic → Schematic capture → Rat's nest → Routing → Artwork → PCB development & etching → Component mounting → Testing

Chapter 4

Network Analysis

A network, in the context of electronics, is a collection of interconnected components. Network analysis is the process of finding the voltages across, and the currents through, every component in the network. There are many different techniques for calculating these values. However, for the most part, the applied technique assumes that the components of the network are all linear. The methods described in this article are only applicable to *linear* network analysis, except where explicitly stated.

Definitions

Component	A device with two or more terminals into which, or out of which, charge may flow.
Node	A point at which terminals of more than two components are joined. A conductor with a substantially zero resistance is considered to be a node for the purpose of analysis.
Branch	The component(s) joining two nodes.
Mesh	A group of branches within a network joined so as to form a complete loop.
Port	Two terminals where the current into one is identical to the current out of the other.
Circuit	A current from one terminal of a generator, through load component(s) and back into the other terminal. A circuit is, in this sense, a one-port network and is a trivial case to analyse. If there is any connection to any other circuits then a non-trivial network has been formed and at least two ports must exist.

Contd...

	Often, "circuit" and "network" are used interchangeably, but many analysts reserve "network" to mean an idealised model consisting of ideal components.
Transfer function	The relationship of the currents and/or voltages between two ports. Most often, an input port and an output port are discussed and the transfer function is described as gain or attenuation.
Component transfer function	For a two-terminal component (i.e. one-port component), the current and voltage are taken as the input and output and the transfer function will have units of impedance or admittance (it is usually a matter of arbitrary convenience whether voltage or current is considered the input). A three (or more) terminal component effectively has two (or more) ports and the transfer function cannot be expressed as a single impedance. The usual approach is to express the transfer function as a matrix of parameters. These parameters can be impedances, but there is a large number of other approaches.

Equivalent Circuits

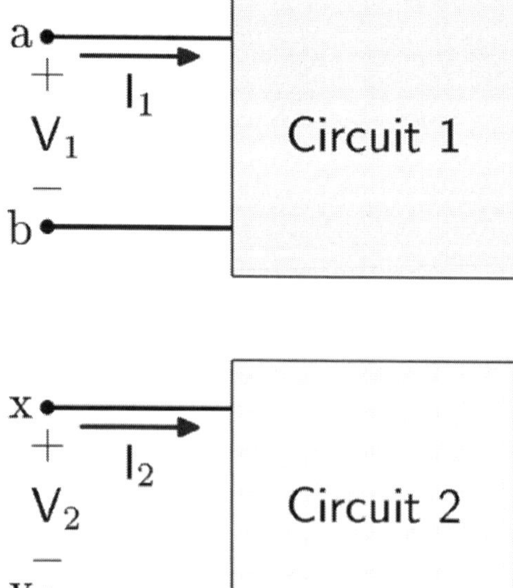

A useful procedure in network analysis is to simplify the network by reducing the number of components. This can be done by replacing

the actual components with other notional components that have the same effect. A particular technique might directly reduce the number of components, for instance by combining impedances in series.

On the other hand it might merely change the form into one in which the components can be reduced in a later operation. For instance, one might transform a voltage generator into a current generator using Norton's theorem in order to be able to later combine the internal resistance of the generator with a parallel impedance load.

A resistive circuit is a circuit containing only resistors, ideal current sources, and ideal voltage sources. If the sources are constant (DC) sources, the result is a DC circuit. The analysis of a circuit refers to the process of solving for the voltages and currents present in the circuit. The solution principles outlined here also apply to phasor analysis of AC circuits.

Two circuits are said to be equivalent with respect to a pair of terminals if the voltage across the terminals and current through the terminals for one network have the same relationship as the voltage and current at the terminals of the other network.

If $V_2 = V_1$ implies $I_2 = I_1$ for all (real) values of V_1, then with respect to terminals ab and xy, circuit 1 and circuit 2 are equivalent.

The above is a sufficient definition for a one-port network. For more than one port, then it must be defined that the currents and voltages between all pairs of corresponding ports must bear the same relationship. For instance, star and delta networks are effectively three port networks and hence require three simultaneous equations to fully specify their equivalence.

Impedances in Series and in Parallel

Any two terminal network of impedances can eventually be reduced to a single impedance by successive applications of impedances in series or impedances in parallel.

Impedances in series: $Z_{eq} = Z_1 + Z_2 + \cdots + Z_n$.

Impedances in parallel: $\dfrac{1}{Z_{eq}} = \dfrac{1}{Z_1} + \dfrac{1}{Z_2} + \cdots + \dfrac{1}{Z_n}$.

The above simplified for only two impedances in parallel:

$$Z_{eq} = \frac{Z_1 Z_2}{Z_1 + Z_2}.$$

Delta-wye Transformation

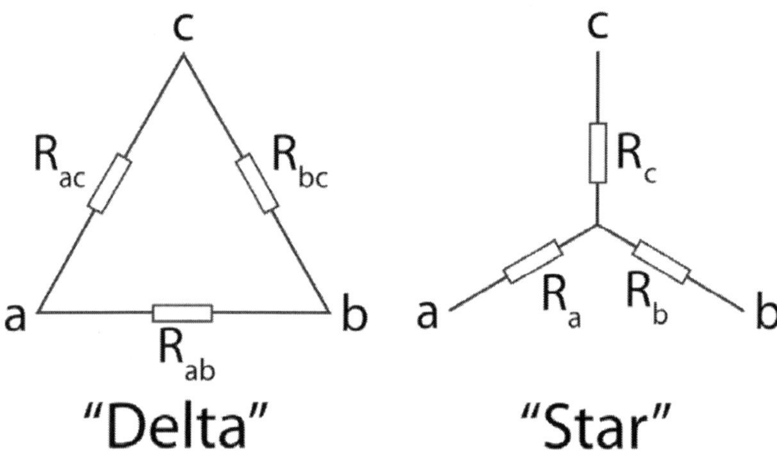

A network of impedances with more than two terminals cannot be reduced to a single impedance equivalent circuit. An n-terminal network can, at best, be reduced to n impedances (at worst ${}^{n}C_{2}$). For a three terminal network, the three impedances can be expressed as a three node delta (D) network or four node star (Y) network.

These two networks are equivalent and the transformations between them are given below. A general network with an arbitrary number of nodes cannot be reduced to the minimum number of impedances using only series and parallel combinations. In general, Y-D and D-Y transformations must also be used.

For some networks the extension of Y-D to star-polygon transformations may also be required.

For equivalence, the impedances between any pair of terminals must be the same for both networks, resulting in a set of three simultaneous equations.

The equations below are expressed as resistances but apply equally to the general case with impedances.

Delta-to-star Transformation Equations

$$R_a = \frac{R_{ac}R_{ab}}{R_{ac} + R_{ab} + R_{bc}}$$

$$R_b = \frac{R_{ab}R_{bc}}{R_{ac} + R_{ab} + R_{bc}}$$

$$R_c = \frac{R_{bc}R_{ac}}{R_{ac} + R_{ab} + R_{bc}}$$

Star-to-delta Transformation Equations

$$R_{ac} = \frac{R_a R_b + R_b R_c + R_c R_a}{R_b}$$

$$R_{ab} = \frac{R_a R_b + R_b R_c + R_c R_a}{R_c}$$

$$R_{bc} = \frac{R_a R_b + R_b R_c + R_c R_a}{R_a}$$

General form of Network Node Elimination

The star-to-delta and series-resistor transformations are special cases of the general resistor network node elimination algorithm. Any node connected by N resistors ($R_1..R_N$) to nodes 1 .. N can be replaced by $\binom{N}{2}$ resistors interconnecting the remaining N nodes. The resistance between x any y two nodes and is given by:

$$R_{xy} = R_x R_y \sum_{i=1}^{N} \frac{1}{R_i}$$

For a star-to-delta ($N = 3$) this reduces to:

$$R_{ab} = R_a R_b \left(\frac{1}{R_a} + \frac{1}{R_b} + \frac{1}{R_c}\right) = \frac{R_a R_b (R_a R_b + R_a R_c + R_b R_c)}{R_a R_b R_c} = \frac{R_a R_b + R_b R_c + R_c R_a}{R_c}$$

For a series reduction ($N = 2$) this reduces to:

$$R_{ab} = R_a R_b \left(\frac{1}{R_a} + \frac{1}{R_b}\right) = \frac{R_a R_b (R_a + R_b)}{R_a R_b} = R_a + R_b$$

For a dangling resistor ($N = 1$) it results in the elimination of the resistor because $\binom{1}{2} = 0$.

Source transformation

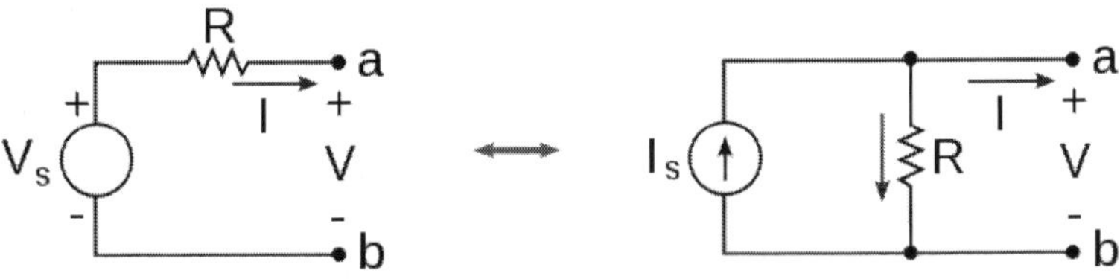

A generator with an internal impedance (i.e. non-ideal generator) can be represented as either an ideal voltage generator or an ideal

current generator plus the impedance. These two forms are equivalent and the transformations are given below. If the two networks are equivalent with respect to terminals ab, then V and I must be identical for both networks. Thus,

$$V_s = RI_s \text{ or } I_s = \frac{V_s}{R}$$

- Norton's theorem states that any two-terminal network can be reduced to an ideal current generator and a parallel impedance.
- Thévenin's theorem states that any two-terminal network can be reduced to an ideal voltage generator plus a series impedance.

Simple Networks

Some very simple networks can be analysed without the need to apply the more systematic approaches.

Voltage Division of Series Components

Consider n impedances that are connected in series. The voltage V_i across any impedance Z_i is

$$V_i = Z_i I = \left(\frac{Z_i}{Z_1 + Z_2 + \cdots + Z_n}\right) V$$

Current Division of Parallel Components

Consider n impedances that are connected in parallel. The current I_i through any impedance Z_i is

$$I_i = \left(\frac{\left(\frac{1}{Z_i}\right)}{\left(\frac{1}{Z_1}\right) + \left(\frac{1}{Z_2}\right) + \cdots + \left(\frac{1}{Z_n}\right)}\right) I$$

for $i = 1, 2, \ldots, n$.

Special case: Current Division of two Parallel Components

$$I_1 = \left(\frac{Z_2}{Z_1 + Z_2}\right) I$$

$$I_2 = \left(\frac{Z_1}{Z_1 + Z_2}\right) I$$

Nodal Analysis

1. Label all nodes in the circuit. Arbitrarily select any node as reference.

2. Define a voltage variable from every remaining node to the reference. These voltage variables must be defined as voltage rises with respect to the reference node.
3. Write a KCL equation for every node except the reference.
4. Solve the resulting system of equations.

Mesh Analysis

Mesh — a loop that does not contain an inner loop.

1. Count the number of "window panes" in the circuit. Assign a mesh current to each window pane.
2. Write a KVL equation for every mesh whose current is unknown.
3. Solve the resulting equations

Superposition

In this method, the effect of each generator in turn is calculated. All the generators other than the one being considered are removed; either short-circuited in the case of voltage generators, or open circuited in the case of current generators. The total current through, or the total voltage across, a particular branch is then calculated by summing all the individual currents or voltages.

There is an underlying assumption to this method that the total current or voltage is a linear superposition of its parts. The method cannot, therefore, be used if non-linear components are present. Note that mesh analysis and node analysis also implicitly use superposition so these too, are only applicable to linear circuits. Superposition cannot be used to find total power consumed by elements even in linear circuits. Power varies according to the square of total voltage (or current) and the square of the sum is not generally equal to sum of the squares.

Choice of Method

Choice of method is to some extent a matter of taste. If the network is particularly simple or only a specific current or voltage is required then ad-hoc application of some simple equivalent circuits may yield the answer without recourse to the more systematic methods.

- Superposition is possibly the most conceptually simple method but rapidly leads to a large number of equations and messy impedance combinations as the network becomes larger.
- Nodal analysis: The number of voltage variables, and hence simultaneous equations to solve, equals the number of nodes minus one. Every voltage source connected to the reference node reduces the number of unknowns (and equations) by one.

- Mesh analysis: The number of current variables, and hence simultaneous equations to solve, equals the number of meshes. Every current source in a mesh reduces the number of unknowns by one. Mesh analysis can only be used with networks which can be drawn as a planar network, that is, with no crossing components.

Transfer Function

A transfer function expresses the relationship between an input and an output of a network. For resistive networks, this will always be a simple real number or an expression which boils down to a real number. Resistive networks are represented by a system of simultaneous algebraic equations. However in the general case of linear networks, the network is represented by a system of simultaneous linear differential equations. In network analysis, rather than use the differential equations directly, it is usual practice to carry out a Laplace transform on them first and then express the result in terms of the Laplace parameter s, which in general is complex. This is described as working in the s-domain. Working with the equations directly would be described as working in the time (or t) domain because the results would be expressed as time varying quantities. The Laplace transform is the mathematical method of transforming between the s-domain and the t-domain.

This approach is standard in control theory and is useful for determining stability of a system, for instance, in an amplifier with feedback.

Two Terminal Component Transfer Functions

For two terminal components the transfer function, or more generally for non-linear elements, the constitutive equation, is the relationship between the current input to the device and the resulting voltage across it. The transfer function, Z(s), will thus have units of impedance – ohms. For the three passive components found in electrical networks, the transfer functions are;

Resistor $Z(s) = R$

Inductor $Z(s) = sL$

Capacitor $Z(s) = \dfrac{1}{sC}$

For a network to which only steady ac signals are applied, s is replaced with $j\omega$ and the more familiar values from ac network theory result.

Resistor $Z(j\omega) = R$

Inductor $Z(j\omega) = j\omega L$

Capacitor $Z(j\omega) = \frac{1}{j\omega C}$

Finally, for a network to which only steady dc is applied, s is replaced with zero and dc network theory applies.

Resistor $Z = R$

Inductor $Z = 0$

Capacitor $Z = \infty$

Two Port Network Transfer Function

Transfer functions, in general, in control theory are given the symbol H(s). Most commonly in electronics, transfer function is defined as the ratio of output voltage to input voltage and given the symbol A(s), or more commonly (because analysis is invariably done in terms of sine wave response), $A(j\omega)$, so that;

$$A(j\omega) = \frac{V_o}{V_i}$$

The A standing for attenuation, or amplification, depending on context. In general, this will be a complex function of $j\omega$, which can be derived from an analysis of the impedances in the network and their individual transfer functions. Sometimes the analyst is only interested in the magnitude of the gain and not the phase angle. In this case the complex numbers can be eliminated from the transfer function and it might then be written as;

$$A(\omega) = \left|\frac{V_o}{V_i}\right|$$

Two Port Parameters

The concept of a two-port network can be useful in network analysis as a black box approach to analysis. The behaviour of the two-port network in a larger network can be entirely characterised without necessarily stating anything about the internal structure. However, to do this it is necessary to have more information than just the A(jω) described above. It can be shown that four such parameters are required to fully characterise the two-port network. These could be the forward transfer function, the input impedance, the reverse transfer function (i.e., the voltage appearing at the input when a voltage is applied to the output) and the output impedance. There are many others, one of these expresses all four parameters as impedances. It is usual to express the four parameters as a matrix;

$$\begin{bmatrix} V_1 \\ V_0 \end{bmatrix} = \begin{bmatrix} z(j\omega)_{11} & z(j\omega)_{12} \\ z(j\omega)_{21} & z(j\omega)_{22} \end{bmatrix} \begin{bmatrix} I_1 \\ I_0 \end{bmatrix}$$

The matrix may be abbreviated to a representative element;

$$[z(j\omega)] \text{ or just } [z]$$

These concepts are capable of being extended to networks of more than two ports. However, this is rarely done in reality because, in many practical cases, ports are considered either purely input or purely output. If reverse direction transfer functions are ignored, a multi-port network can always be decomposed into a number of two-port networks.

Distributed Components

Where a network is composed of discrete components, analysis using two-port networks is a matter of choice, not essential. The network can always alternatively be analysed in terms of its individual component transfer functions. However, if a network contains distributed components, such as in the case of a transmission line, then it is not possible to analyse in terms of individual components since they do not exist.

The most common approach to this is to model the line as a two-port network and characterise it using two-port parameters (or something equivalent to them). Another example of this technique is modelling the carriers crossing the base region in a high frequency transistor. The base region has to be modelled as distributed resistance and capacitance rather than lumped components.

Image Analysis

Transmission lines and certain types of filter design use the image method to determine their transfer parameters. In this method, the behaviour of an infinitely long cascade connected chain of identical networks is considered. The input and output impedances and the forward and reverse transmission functions are then calculated for this infinitely long chain. Although the theoretical values so obtained can never be exactly realised in practice, in many cases they serve as a very good approximation for the behaviour of a finite chain as long as it is not too short.

Non-linear Networks

Most electronic designs are, in reality, non-linear. There is very little that does not include some semiconductor devices. These are invariably non-linear, the transfer function of an ideal semiconductor p-n junction is given by the very non-linear relationship;

$$i = I_o(e^{\frac{v}{V_T}} - 1)$$

where;

- i and v are the instantaneous current and voltage.
- I_o is an arbitrary parameter called the reverse leakage current whose value depends on the construction of the device.
- V_T is a parameter proportional to temperature called the thermal voltage and equal to about 25mV at room temperature.

There are many other ways that non-linearity can appear in a network. All methods utilising linear superposition will fail when non-linear components are present. There are several options for dealing with non-linearity depending on the type of circuit and the information the analyst wishes to obtain.

Constitutive Equations

The diode equation above is an example of an element constitutive equation of the general form,

$$f(v,i) = 0$$

This can be thought of as a non-linear resistor. The corresponding constitutive equations for non-linear inductors and capacitors are respectively;

$$f(v,\varphi) = 0$$

$$f(v,q) = 0$$

where f is any arbitrary function, φ is the stored magnetic flux and q is the stored charge.

Existence, Uniqueness and Stability

An important consideration in non-linear analysis is the question of uniqueness. For a network composed of linear components there will always be one, and only one, unique solution for a given set of boundary conditions. This is not always the case in non-linear circuits. For instance, a linear resistor with a fixed current applied to it has only one solution for the voltage across it. On the other hand, the non-linear tunnel diode has up to three solutions for the voltage for a given current. That is, a particular solution for the current through the diode is not unique, there may be others, equally valid. In some cases there may not be a solution at all: the question of existence of solutions must be considered.

Another important consideration is the question of stability. A particular solution may exist, but it may not be stable, rapidly departing from that point at the slightest stimulation. It can be shown that a network that is absolutely stable for all conditions must have one, and only one, solution for each set of conditions.

Methods

Boolean Analysis of Switching Networks

A switching device is one where the non-linearity is utilised to produce two opposite states. CMOS devices in digital circuits, for instance, have their output connected to either the positive or the negative supply rail and are never found at anything in between except during a transient period when the device is actually switching. Here the non-linearity is designed to be extreme, and the analyst can actually take advantage of that fact.

These kinds of networks can be analysed using Boolean algebra by assigning the two states ("on"/"off", "positive"/"negative" or whatever states are being used) to the boolean constants "0" and "1".

The transients are ignored in this analysis, along with any slight discrepancy between the actual state of the device and the nominal state assigned to a boolean value. For instance, boolean "1" may be assigned to the state of +5V.

The output of the device may actually be +4.5V but the analyst still considers this to be boolean "1". Device manufacturers will usually specify a range of values in their data sheets that are to be considered undefined (i.e. the result will be unpredictable).

The transients are not entirely uninteresting to the analyst. The maximum rate of switching is determined by the speed of transition from one state to the other. Happily for the analyst, for many devices most of the transition occurs in the linear portion of the devices transfer function and linear analysis can be applied to obtain at least an approximate answer.

It is mathematically possible to derive boolean algebras which have more than two states. There is not too much use found for these in electronics, although three-state devices are passingly common.

Separation of Bias and Signal Analyses

This technique is used where the operation of the circuit is to be essentially linear, but the devices used to implement it are non-linear. A transistor amplifier is an example of this kind of network. The essence of this technique is to separate the analysis into two parts. Firstly, the dc biases are analysed using some non-linear method. This establishes the quiescent operating point of the circuit. Secondly, the small signal characteristics of the circuit are analysed using linear network analysis. Examples of methods that can be used for both these stages are given below.

Graphical Method of dc Analysis

In a great many circuit designs, the dc bias is fed to a non-linear component via a resistor (or possibly a network of resistors). Since resistors are linear components, it is particularly easy to determine the quiescent operating point of the non-linear device from a graph of its transfer function. The method is as follows: from linear network analysis the output transfer function (that is output voltage against output current) is calculated for the network of resistor(s) and the generator driving them. This will be a straight line (called the load line) and can readily be superimposed on the transfer function plot of the non-linear device. The point where the lines cross is the quiescent operating point.

Perhaps the easiest practical method is to calculate the (linear) network open circuit voltage and short circuit current and plot these on the transfer function of the non-linear device. The straight line joining these two point is the transfer function of the network.

In reality, the designer of the circuit would proceed in the reverse direction to that described. Starting from a plot provided in the manufacturers data sheet for the non-linear device, the designer would choose the desired operating point and then calculate the linear component values required to achieve it.

It is still possible to use this method if the device being biased has its bias fed through another device which is itself non-linear – a diode for instance. In this case however, the plot of the network transfer function onto the device being biased would no longer be a straight line and is consequently more tedious to do.

Small Signal Equivalent Circuit

This method can be used where the deviation of the input and output signals in a network stay within a substantially linear portion of the non-linear devices transfer function, or else are so small that the curve of the transfer function can be considered linear. Under a set of these specific conditions, the non-linear device can be represented by an equivalent linear network. It must be remembered that this equivalent circuit is entirely notional and only valid for the small signal deviations. It is entirely inapplicable to the dc biasing of the device.

For a simple two-terminal device, the small signal equivalent circuit may be no more than two components. A resistance equal to the slope of the v/i curve at the operating point (called the dynamic resistance), and tangent to the curve. A generator, because this tangent will not, in general, pass through the origin. With more terminals, more complicated equivalent circuits are required.

A popular form of specifying the small signal equivalent circuit amongst transistor manufacturers is to use the two-port network parameters known as [h] parameters. These are a matrix of four parameters as with the [z] parameters but in the case of the [h] parameters they are a hybrid mixture of impedances, admittances, current gains and voltage gains. In this model the three terminal transistor is considered to be a two port network, one of its terminals being common to both ports. The [h] parameters are quite different depending on which terminal is chosen as the common one. The most important parameter for transistors is usually the forward current gain, h_{21}, in the common emitter configuration. This is designated h_{fe} on data sheets.

The small signal equivalent circuit in terms of two-port parameters leads to the concept of dependent generators. That is, the value of a voltage or current generator depends linearly on a voltage or current elsewhere in the circuit. For instance the [z] parameter model leads to dependent voltage generators as shown in this diagram;

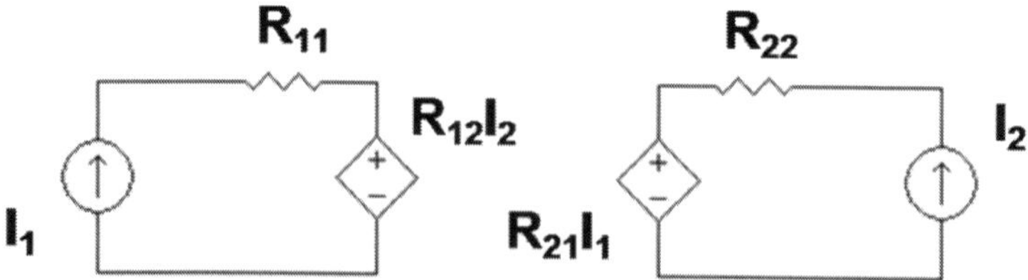

Figure: *[z] parameter equivalent circuit showing dependent voltage generators*

There will always be dependent generators in a two-port parameter equivalent circuit. This applies to the [h] parameters as well as to the

[z] and any other kind. These dependencies must be preserved when developing the equations in a larger linear network analysis.

Piecewise Linear Method

In this method, the transfer function of the non-linear device is broken up into regions. Each of these regions is approximated by a straight line. Thus, the transfer function will be linear up to a particular point where there will be a discontinuity. Past this point the transfer function will again be linear but with a different slope.

A well known application of this method is the approximation of the transfer function of a pn junction diode. The actual transfer function of an ideal diode has been given at the top of this (non-linear) section. However, this formula is rarely used in network analysis, a piecewise approximation being used instead. It can be seen that the diode current rapidly diminishes to $-I_0$ as the voltage falls. This current, for most purposes, is so small it can be ignored. With increasing voltage, the current increases exponentially. The diode is modelled as an open circuit up to the knee of the exponential curve, then past this point as a resistor equal to the bulk resistance of the semiconducting material.

The commonly accepted values for the transition point voltage are 0.7V for silicon devices and 0.3V for germanium devices. An even simpler model of the diode, sometimes used in switching applications, is short circuit for forward voltages and open circuit for reverse voltages.

The model of a forward biased pn junction having an approximately constant 0.7V is also a much used approximation for transistor base-emitter junction voltage in amplifier design.

The piecewise method is similar to the small signal method in that linear network analysis techniques can only be applied if the signal stays within certain bounds. If the signal crosses a discontinuity point then the model is no longer valid for linear analysis purposes. The model does have the advantage over small signal however, in that it is equally applicable to signal and dc bias. These can therefore both be analysed in the same operations and will be linearly superimposable.

Time-varying Components

In linear analysis, the components of the network are assumed to be unchanging, but in some circuits this does not apply, such as sweep oscillators, voltage controlled amplifiers, and variable equalisers. In many circumstances the change in component value is periodic. A non-linear component excited with a periodic signal, for instance, can be represented as periodically varying *linear* component. Sidney

Darlington disclosed a method of analysing such periodic time varying circuits. He developed canonical circuit forms which are analogous to the canonical forms of Ronald Foster and Wilhelm Cauer used for analysing linear circuits.

LC Circuit

An LC circuit, also called a resonant circuit, tank circuit, or tuned circuit, is an electric circuit consisting of an inductor, represented by the letter L, and a capacitor, represented by the letter C, connected together. The circuit can act as an electrical resonator, an electrical analogue of a tuning fork, storing energy oscillating at the circuit's resonant frequency. The two-element LC circuit is the simplest type of inductor-capacitor network (or LC network). The two-element circuit is also referred to as a *second order LC circuit* to distinguish it from more complicated (higher order) LC networks with more inductors and capacitors.

LC circuits are used either for generating signals at a particular frequency, or picking out a signal at a particular frequency from a more complex signal. They are key components in many electronic devices, particularly radio equipment, used in circuits such as oscillators, filters, tuners and frequency mixers.

An LC circuit is an idealized model since it assumes there is no dissipation of energy due to resistance. Any practical implementation of an LC circuit will always include loss resulting from small but non-zero resistance within the components and connecting wires. The purpose of an LC circuit is usually to oscillate with minimal damping, so the resistance is made as low as possible. While no practical circuit is without losses, it is nonetheless instructive to study this ideal form of the circuit to gain understanding and physical intuition.

Operation

An LC circuit can store electrical energy oscillating at its resonant frequency. A capacitor stores energy in the electric field (E) between its plates, depending on the voltage across it, and an inductor stores energy in its magnetic field (B), depending on the current through it.

If a charged capacitor is connected across an inductor, charge will start to flow through the inductor, building up a magnetic field around it and reducing the voltage on the capacitor. Eventually all the charge on the capacitor will be gone and the voltage across it will reach zero. However, the current will continue, because inductors resist changes in current. The energy to keep it flowing is extracted from the magnetic field, which will begin to decline. The current will begin to charge the capacitor with a voltage of opposite polarity to its original charge.

When the magnetic field is completely dissipated the current will stop and the charge will again be stored in the capacitor, with the opposite polarity as before. Then the cycle will begin again, with the current flowing in the opposite direction through the inductor.

The charge flows back and forth between the plates of the capacitor, through the inductor. The energy oscillates back and forth between the capacitor and the inductor until (if not replenished by power from an external circuit) internal resistance makes the oscillations die out. Its action, known mathematically as a harmonic oscillator, is similar to a pendulum swinging back and forth, or water sloshing back and forth in a tank.

For this reason the circuit is also called a tank circuit. The oscillation frequency is determined by the capacitance and inductance values. In typical tuned circuits in electronic equipment the oscillations are very fast, thousands to billions of times per second.

Resonance Effect

The resonance effect occurs when inductive and capacitive reactances are equal in magnitude. The frequency at which this equality holds for the particular circuit is called the resonant frequency. The resonant frequency of the LC circuit is

$$\omega_0 = \frac{1}{\sqrt{LC}}$$

where L is the inductance in henries, and C is the capacitance in farads. The angular frequency ω_0 has units of radians per second.

The equivalent frequency in units of hertz is

$$f_0 = \frac{\omega_0}{2\pi} = \frac{1}{2\pi\sqrt{LC}}.$$

LC circuits are often used as filters; the L/C ratio is one of the factors that determines their "Q" and so selectivity. For a series resonant circuit with a given resistance, the higher the inductance and the lower the capacitance, the narrower the filter bandwidth. For a parallel resonant circuit the opposite applies. Positive feedback around the tuned circuit ("regeneration") can also increase selectivity.

Stagger tuning can provide an acceptably wide audio bandwidth, yet good selectivity.

RC Circuit

A resistor–capacitor circuit (RC circuit), or RC filter or RC network, is an electric circuit composed of resistors and capacitors driven by a voltage or current source. A first order RC circuit is composed of one resistor and one capacitor and is the simplest type of RC circuit.

RC circuits can be used to filter a signal by blocking certain frequencies and passing others. The two most common RC filters are the high-pass filters and low-pass filters; band-pass filters and band-stop filters usually require RLC filters, though crude ones can be made with RC filters.

Introduction

There are three basic, linear passive lumped analog circuit components: the resistor (R), the capacitor (C), and the inductor (L). These may be combined in the RC circuit, the RL circuit, the LC circuit, and the RLC circuit, with the abbreviations indicating which components are used. These circuits, among them, exhibit a large number of important types of behaviour that are fundamental to much of analog electronics. In particular, they are able to act as passive filters. This article considers the RC circuit, in both series and parallel forms, as shown in the diagrams below.

This article relies on knowledge of the complex impedance representation of capacitors and on knowledge of the frequency domain representation of signals.

Natural Response

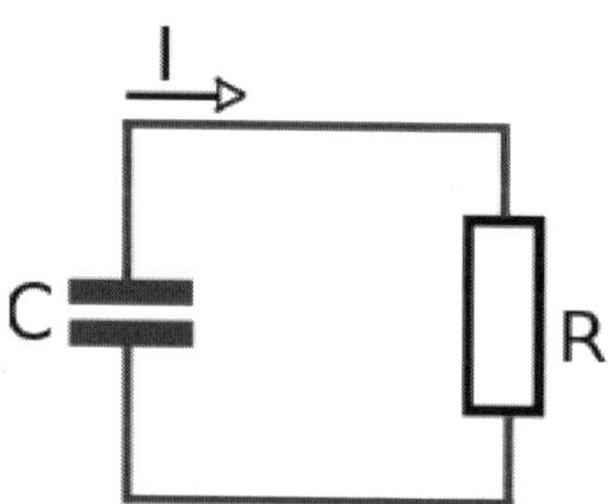

Figure: *RC circuit*

The simplest RC circuit is a capacitor and a resistor in series. When a circuit consists of only a charged capacitor and a resistor, the capacitor will discharge its stored energy through the resistor. The voltage across the capacitor, which is time dependent, can be found by using Kirchhoff's current law, where the current charging the capacitor must equal the current through the resistor. This results in the linear differential equation

$$C\frac{dV}{dt}+\frac{V}{R}=0\,.$$

Solving this equation for V yields the formula for exponential decay:

$$V(t)=V_0 e^{-\frac{t}{RC}}\,,$$

where V_0 is the capacitor voltage at time $t = 0$.

The time required for the voltage to fall to $\frac{V_0}{e}$ is called the RC time constant and is given by

$$\tau = RC\,.$$

Complex Impedance

The complex impedance, Z_C (in ohms) of a capacitor with capacitance C (in farads) is

$$Z_C=\frac{1}{sC}$$

The complex frequency s is, in general, a complex number,

$$s=\sigma + j\omega$$

where

- j represents the imaginary unit:

  $$j^2=-1$$

- σ is the exponential decay constant (in radians per second), and
- ω is the sinusoidal angular frequency (also in radians per second).

Sinusoidal Steady State

Sinusoidal steady state is a special case in which the input voltage consists of a pure sinusoid (with no exponential decay). As a result,

$$\sigma = 0$$

and the evaluation of s becomes

$$s=j\omega$$

Series circuit

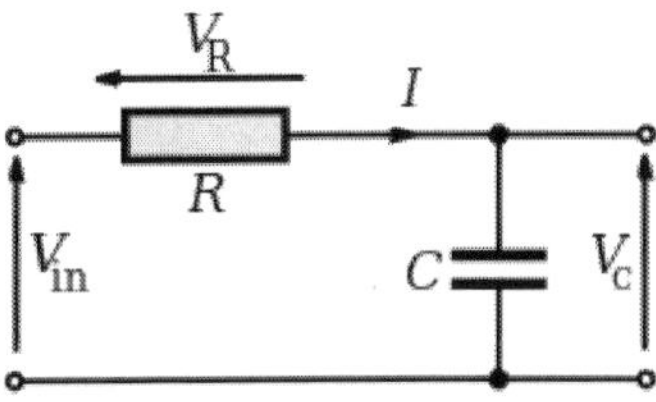

Figure: *Series RC circuit*

By viewing the circuit as a voltage divider, the voltage across the capacitor is:

$$V_C(s) = \frac{1/Cs}{R+1/Cs} V_{in}(s) = \frac{1}{1+RCs} V_{in}(s)$$

and the voltage across the resistor is:

$$V_R(s) = \frac{R}{R+1/Cs} V_{in}(s) = \frac{RCs}{1+RCs} V_{in}(s).$$

Transfer functions

The transfer function from the input voltage to the voltage across the capacitor is

$$H_C(s) = \frac{V_C(s)}{V_{in}(s)} = \frac{1}{1+RCs}.$$

Similarly, the transfer function from the input to the voltage across the resistor is

$$H_R(s) = \frac{V_R(s)}{V_{in}(s)} = \frac{RCs}{1+RCs}.$$

Poles and Zeros

Both transfer functions have a single pole located at

$$s = -\frac{1}{RC}.$$

In addition, the transfer function for the resistor has a zero located at the origin.

Gain and Phase

The magnitude of the gains across the two components are:

$$G_C = |H_C(j\omega)| = \left|\frac{V_C(j\omega)}{V_{in}(j\omega)}\right| = \frac{1}{\sqrt{1+(\omega RC)^2}}$$

and

$$G_R = |H_R(j\omega)| = \left|\frac{V_R(j\omega)}{V_{in}(j\omega)}\right| = \frac{\omega RC}{\sqrt{1+(\omega RC)^2}},$$

and the phase angles are:

$$\phi_C = \angle H_C(j\omega) = \tan^{-1}(-\omega RC)$$

and

$$\phi_R = \angle H_R(j\omega) = \tan^{-1}\left(\frac{1}{\omega RC}\right).$$

These expressions together may be substituted into the usual expression for the phasor representing the output:

$$V_C = G_C V_{in} e^{j\phi C}$$

$$V_R = G_R V_{in} e^{j\phi R}.$$

Current

The current in the circuit is the same everywhere since the circuit is in series:

$$I(s) = \frac{V_{in}(s)}{R + \frac{1}{Cs}} = \frac{Cs}{1 + RCs} V_{in}(s)$$

Impulse response

The impulse response for each voltage is the inverse Laplace transform of the corresponding transfer function. It represents the response of the circuit to an input voltage consisting of an impulse or Dirac delta function. The impulse response for the capacitor voltage is

$$h_C(t) = \frac{1}{RC} e^{-t/RC} u(t) = \frac{1}{\tau} e^{-t/\tau} u(t)$$

where $u(t)$ is the Heaviside step function and

$$\tau = RC$$

is the time constant. Similarly, the impulse response for the resistor voltage is

$$h_R(t) = \delta(t) - \frac{1}{RC} e^{-t/RC} u(t) = \delta(t) - \frac{1}{\tau} e^{-t/\tau} u(t)$$

where $\delta(t)$ is the Dirac delta function

Frequency-domain Considerations

These are frequency domain expressions. Analysis of them will show which frequencies the circuits (or filters) pass and reject. This analysis rests on a consideration of what happens to these gains as the frequency becomes very large and very small.

As $\omega \to \infty$:

$$G_C \to 0$$

$$G_R \to 1.$$

As $\omega \to 0$:

$$G_C \to 1$$

$$G_R \to 0.$$

This shows that, if the output is taken across the capacitor, high frequencies are attenuated (shorted to ground) and low frequencies are passed. Thus, the circuit behaves as a *low-pass filter*. If, though, the output is taken across the resistor, high frequencies are passed and low frequencies are attenuated (since the capacitor blocks the signal as its frequency approaches 0). In this configuration, the circuit behaves as a *high-pass filter*.

The range of frequencies that the filter passes is called its bandwidth. The point at which the filter attenuates the signal to half its unfiltered power is termed its cutoff frequency. This requires that the gain of the circuit be reduced to

$$G_C = G_R = \frac{1}{\sqrt{2}}.$$

Solving the above equation yields

$$\omega_c = \frac{1}{RC}$$

or

$$f_c = \frac{1}{2\pi RC}$$

which is the frequency that the filter will attenuate to half its original power.

Clearly, the phases also depend on frequency, although this effect is less interesting generally than the gain variations.

As $\omega \to 0$:

$$\phi_C \to 0$$

$$\phi_R \to 90^\circ = \pi / 2^c.$$

As $\omega \to \infty$:

$$\phi_C \to -90^\circ = -\pi / 2^c$$

$$\phi_R \to 0$$

So at DC (0 Hz), the capacitor voltage is in phase with the signal voltage while the resistor voltage leads it by 90°. As frequency increases, the capacitor voltage comes to have a 90° lag relative to the signal and the resistor voltage comes to be in-phase with the signal.

Time-domain Considerations

The most straightforward way to derive the time domain behaviour is to use the Laplace transforms of the expressions for V_C and V_R given above. This effectively transforms $j\omega \rightarrow s$. Assuming a step input (i.e. $V_{in} = 0$ before $t = 0$ and then $V_{in} = V$ afterwards):

$$V_{in}(s) = V\frac{1}{s}$$

$$V_C(s) = V\frac{1}{1+sRC}\frac{1}{s}$$

and

$$V_R(s) = V\frac{sRC}{1+sRC}\frac{1}{s}.$$

Figure: *Capacitor voltage step-response.*

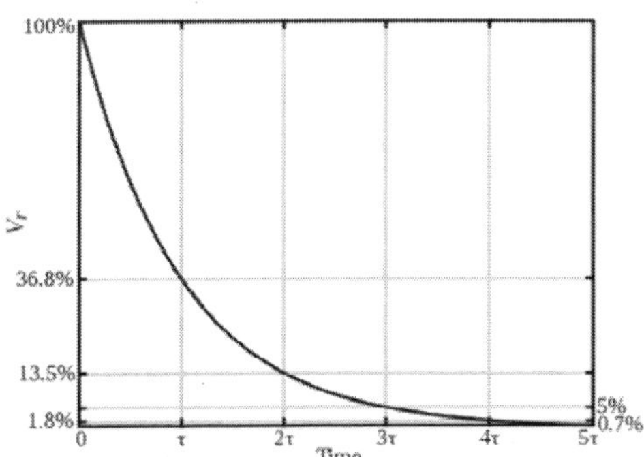

Figure: *Resistor voltage step-response.*

Partial fractions expansions and the inverse Laplace transform yield:

$$V_C(t) = V\left(1 - e^{-t/RC}\right)$$

$$V_R(t) = Ve^{-t/RC}.$$

These equations are for calculating the voltage across the capacitor and resistor respectively while the capacitor is charging; for discharging, the equations are vice-versa. These equations can be rewritten in terms of charge and current using the relationships C=Q/V and V=IR.

Thus, the voltage across the capacitor tends towards V as time passes, while the voltage across the resistor tends towards 0, as shown in the figures. This is in keeping with the intuitive point that the capacitor will be charging from the supply voltage as time passes, and will eventually be fully charged.

These equations show that a series RC circuit has a time constant, usually denoted $\tau = RC$ being the time it takes the voltage across the component to either rise (across C) or fall (across R) to within $1/e$ of its final value. That is, τ is the time it takes V_C to reach $V(1-1/e)$ and V_R to reach $V(1/e)$.

The rate of change is a *fractional* $\left(1-\frac{1}{e}\right)$ per τ. Thus, in going from $t = N\tau$ to $t = (N+1)\tau$, the voltage will have moved about 63.2% of the way from its level at $t = N\tau$ toward its final value. So C will be charged to about 63.2% after τ, and essentially fully charged (99.3%) after about 5τ. When the voltage source is replaced with a short-circuit, with C fully charged, the voltage across C drops exponentially with t from V towards 0. C will be discharged to about 36.8% after τ, and essentially fully discharged (0.7%) after about 5τ. Note that the current, I, in the circuit behaves as the voltage across R does, via Ohm's Law.

These results may also be derived by solving the differential equations describing the circuit:

$$\frac{V_{in} - V_C}{R} = C\frac{dV_C}{dt}$$

and

$$V_R = V_{in} - V_C.$$

The first equation is solved by using an integrating factor and the second follows easily; the solutions are exactly the same as those obtained via Laplace transforms.

Integrator

Consider the output across the capacitor at *high* frequency i.e.

$$\omega \gg \frac{1}{RC}.$$

This means that the capacitor has insufficient time to charge up and so its voltage is very small. Thus the input voltage approximately equals the voltage across the resistor. To see this, consider the expression for I given above:

$$I = \frac{V_{in}}{R + 1/j\omega C}$$

but note that the frequency condition described means that

$$\omega C \gg \frac{1}{R}$$

so

$$I \approx \frac{V_{in}}{R} \text{ which is just Ohm's Law.}$$

Now,

$$V_C = \frac{1}{C}\int_0^t I dt$$

so

$$V_C \approx \frac{1}{RC}\int_0^t V_{in} dt \,,$$

which is an integrator *across the capacitor.*

Differentiator

Consider the output across the resistor at *low* frequency i.e.,

$$\omega \ll \frac{1}{RC}.$$

This means that the capacitor has time to charge up until its voltage is almost equal to the source's voltage. Considering the expression for I again, when

$$R \ll \frac{1}{\omega C},$$

so

$$I \approx \frac{V_{in}}{1/j\omega C}$$

$$V_{in} \approx \frac{I}{j\omega C} = V_C$$

Now,

$$V_R = IR = C\frac{dV_C}{dt}R$$

$$V_R \approx RC\frac{dV_{in}}{dt}$$

which is a differentiator *across the resistor.*

More accurate integration and differentiation can be achieved by placing resistors and capacitors as appropriate on the input and feedback loop of operational amplifiers.

Parallel Circuit

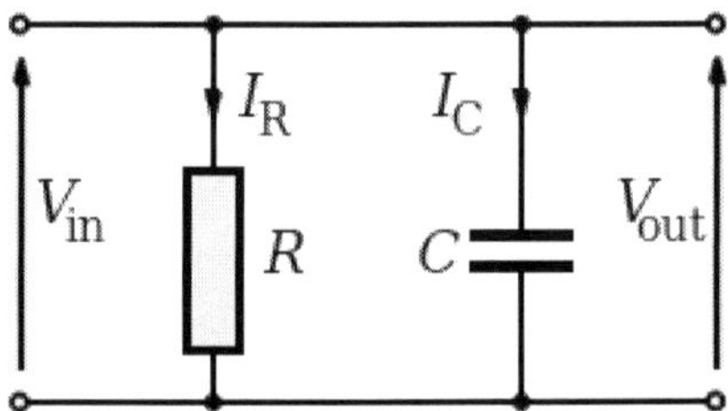

Figure: *Parallel RC circuit*

The parallel RC circuit is generally of less interest than the series circuit. This is largely because the output voltage V_{out} is equal to the input voltage V_{in} — as a result, this circuit does not act as a filter on the input signal unless fed by a current source.

With complex impedances:

$$I_R = \frac{V_{in}}{R}$$

and

$$I_C = j\omega CV_{in}\ .$$

This shows that the capacitor current is 90° out of phase with the resistor (and source) current. Alternatively, the governing differential equations may be used:

$$I_R = \frac{V_{in}}{R}$$

and

$$I_C = C\frac{dV_{in}}{dt}.$$

When fed by a current source, the transfer function of a parallel RC circuit is:

$$\frac{V_{out}}{I_{in}} = \frac{R}{1+sRC}.$$

RL Circuit

A resistor–inductor circuit (RL circuit), or RL filter or RL network, is an electric circuit composed of resistors and inductors driven by a voltage or current source. A first order RL circuit is composed of one resistor and one inductor and is the simplest type of RL circuit.

A first order RL circuit is one of the simplest analogue infinite impulse response electronic filters. It consists of a resistor and an inductor, either in series driven by a voltage source or in parallel driven by a current source.

Introduction

The fundamental passive linear circuit elements are the resistor (R), capacitor (C) and inductor (L). These circuit elements can be combined to form an electrical circuit in four distinct ways: the RC circuit, the RL circuit, the LC circuit and the RLC circuit with the abbreviations indicating which components are used. These circuits exhibit important types of behaviour that are fundamental to analogue electronics. In particular, they are able to act as passive filters. This article considers the RL circuit in both series and parallel as shown in the diagrams.

In practice, however, capacitors (and RC circuits) are usually preferred to inductors since they can be more easily manufactured and are generally physically smaller, particularly for higher values of components.

Both RC and RL circuits form a single-pole filter. Depending on whether the reactive element (C or L) is in series with the load, or parallel with the load will dictate whether the filter is low-pass or high-pass.

Frequently RL circuits are used for DC power supplies to RF amplifiers, where the inductor is used to pass DC bias current and block the RF getting back into the power supply.

Complex Impedance

The complex impedance Z_L (in ohms) of an inductor with inductance L (in henries) is

$$Z_L = Ls$$

The complex frequency s is a complex number,

$$s = \sigma + j\omega$$

where

- j represents the imaginary unit:

$$j^2 = -1$$

- σ is the exponential decay constant (in radians per second), and
- ω is the angular frequency (in radians per second).

EigenFunctions

The complex-valued eigenfunctions of *any* linear time-invariant (LTI) system are of the following forms:

$$\begin{aligned} \mathrm{V}(t) &= \mathrm{A}e^{st} = \mathrm{A}e^{(\sigma+j\omega)t} \\ \mathrm{A} &= Ae^{j\phi} \\ \Rightarrow \mathrm{V}(t) &= Ae^{j\phi}e^{(\sigma+j\omega)t} \\ &= Ae^{\sigma t}e^{j(\omega t+\phi)} \end{aligned}$$

From Euler's formula, the real-part of these eigenfunctions are exponentially-decaying sinusoids:

$$v(t) = \mathrm{Re}\{V(t)\} = Ae^{\sigma t}\cos(\omega t + \phi)$$

Sinusoidal Steady State

Sinusoidal steady state is a special case in which the input voltage consists of a pure sinusoid (with no exponential decay). As a result,

$$\sigma = 0$$

and the evaluation of s becomes

$$s = j\omega$$

Series Circuit

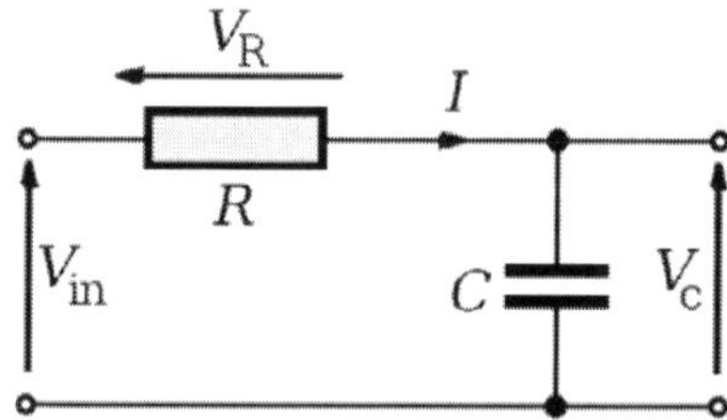

Figure: *Series RL circuit*

By viewing the circuit as a voltage divider, we see that the voltage across the inductor is:

$$V_L(s) = \frac{Ls}{R+Ls}V_{in}(s)$$

and the voltage across the resistor is:

$$V_R(s) = \frac{R}{R+Ls}V_{in}(s)$$

Current

The current in the circuit is the same everywhere since the circuit is in series:

$$I(s) = \frac{V_{in}(s)}{R + Ls}$$

Transfer Functions

The transfer function for the inductor is

$$H_L(s) = \frac{V_L(s)}{V_{in}(s)} = \frac{Ls}{R + Ls} = G_L e^{j\phi L}$$

Similarly, the transfer function for the resistor is

$$H_R(s) = \frac{V_R(s)}{V_{in}(s)} = \frac{R}{R + Ls} = G_R e^{j\phi R}$$

Poles and Zeros

Both transfer functions have a single pole located at

$$s = -\frac{R}{L}$$

In addition, the transfer function for the inductor has a zero located at the origin.

Gain and Phase Angle

The gains across the two components are found by taking the magnitudes of the above expressions:

$$G_L = |H_L(s)| = \left|\frac{V_L(s)}{V_{in}(s)}\right| = \frac{\omega L}{\sqrt{R^2 + (\omega L)^2}}$$

and

$$G_R = |H_R(s)| = \left|\frac{V_R(s)}{V_{in}(s)}\right| = \frac{R}{\sqrt{R^2 + (\omega L)^2}}$$

and the phase angles are:

$$\phi_L = \angle H_L(s) = \tan^{-1}\left(\frac{R}{\omega L}\right)$$

and

$$\phi_R = \angle H_R(s) = \tan^{-1}\left(-\frac{\omega L}{R}\right)$$

Phasor Notation

These expressions together may be substituted into the usual expression for the phasor representing the output:

$$V_L = G_L V_{in} e^{j\phi L}$$
$$V_R = G_R V_{in} e^{j\phi R}$$

Impulse Response

The impulse response for each voltage is the inverse Laplace transform of the corresponding transfer function. It represents the response of the circuit to an input voltage consisting of an impulse or Dirac delta function.

The impulse response for the inductor voltage is

$$h_L(t) = \delta(t) - \frac{R}{L} e^{-t\frac{R}{L}} u(t) = \delta(t) - \frac{1}{\tau} e^{-\frac{1}{\tau}t} u(t)$$

where $u(t)$ is the Heaviside step function and

$$\tau = \frac{L}{R}$$

is the time constant. Similarly, the impulse response for the resistor voltage is

$$h_R(t) = \frac{R}{L} e^{-t\frac{R}{L}} u(t) = \frac{1}{\tau} e^{-\frac{1}{\tau}t} u(t)$$

Zero Input Response (ZIR)

The Zero input response, also called the natural response, of an RL circuit describes the behaviour of the circuit after it has reached constant voltages and currents and is disconnected from any power source. It is called the zero-input response because it requires no input.

The ZIR of an RL circuit is:

$$i(t) = i(0) e^{-\frac{R}{L}t} = i(0) e^{-\frac{1}{\tau}t}$$

Frequency Domain Considerations

These are frequency domain expressions. Analysis of them will show which frequencies the circuits (or filters) pass and reject. This analysis rests on a consideration of what happens to these gains as the frequency becomes very large and very small.

As $\omega \to \infty$:

$$G_L \to 1$$
$$G_R \to 0$$

As $\omega \to 0$:

$$G_L \quad \to 0$$
$$G_R \quad \to 1$$

This shows that, if the output is taken across the inductor, high frequencies are passed and low frequencies are attenuated (rejected). Thus, the circuit behaves as a *high-pass filter*. If, though, the output is taken across the resistor, high frequencies are rejected and low frequencies are passed. In this configuration, the circuit behaves as a *low-pass filter*. Compare this with the behaviour of the resistor output in an RC circuit, where the reverse is the case.

The range of frequencies that the filter passes is called its bandwidth. The point at which the filter attenuates the signal to half its unfiltered power is termed its cutoff frequency. This requires that the gain of the circuit be reduced to

$$G_L = G_R = \frac{1}{\sqrt{2}}.$$

Solving the above equation yields

$$\omega_c = \frac{R}{L}\,\text{rad/s}$$

or

$$f_c = \frac{R}{2\pi L}\,\text{Hz}$$

which is the frequency that the filter will attenuate to half its original power.

Clearly, the phases also depend on frequency, although this effect is less interesting generally than the gain variations.

As $\omega \to 0$:

$$\phi_L \quad \to 90^\circ = \frac{\pi}{2^c}$$
$$\phi_R \quad \to 0$$

As $\omega \to \infty$:

$$\phi_L \quad \to 0$$
$$\phi_R \quad \to -90^\circ = -\frac{\pi}{2^c}$$

So at DC (0 Hz), the resistor voltage is in phase with the signal voltage while the inductor voltage leads it by 90°. As frequency increases, the resistor voltage comes to have a 90° lag relative to the signal and the inductor voltage comes to be in-phase with the signal.

Time Domain Considerations

The most straightforward way to derive the time domain behaviour is to use the Laplace transforms of the expressions for V_L and V_R given above. This effectively transforms $j\omega \to s$. Assuming a step input (i.e., $V_{in} = 0$ before $t = 0$ and then $V_{in} = V$ afterwards):

$$V_{in}(s) = V\frac{1}{s}$$

$$V_L(s) = V\frac{sL}{R+sL}\frac{1}{s}$$

$$V_R(s) = V\frac{R}{R+sL}\frac{1}{s}$$

Figure: *Inductor voltage step-response.*

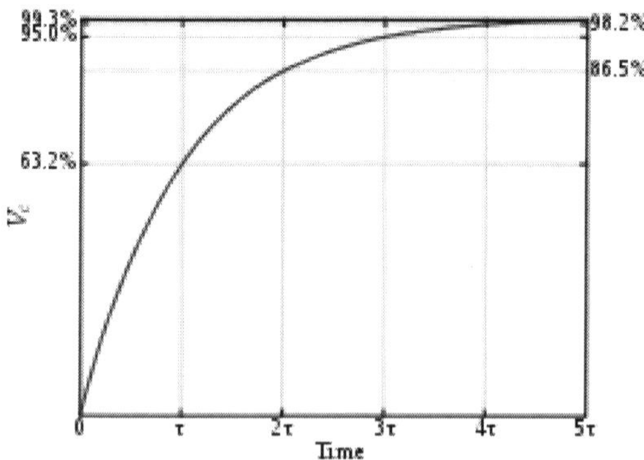

Figure: *Resistor voltage step-response.*

Partial fractions expansions and the inverse Laplace transform yield:

$$V_L(t) = Ve^{-t\frac{R}{L}}$$

$$V_R(t) = V\left(1 - e^{-t\frac{R}{L}}\right)$$

Thus, the voltage across the inductor tends towards 0 as time passes, while the voltage across the resistor tends towards V, as shown in the figures. This is in keeping with the intuitive point that the inductor will only have a voltage across as long as the current in the circuit is changing — as the circuit reaches its steady-state, there is no further current change and ultimately no inductor voltage.

These equations show that a series RL circuit has a time constant, usually denoted $\tau = \frac{L}{R}$ being the time it takes the voltage across the component to either fall (across L) or rise (across R) to within $\frac{1}{e}$ of its final value. That is, τ is the time it takes V_L to reach $V\left(\frac{1}{e}\right)$ and V_R to reach $V\left(1 - \frac{1}{e}\right)$.

The rate of change is a *fractional* $\left(1 - \frac{1}{e}\right)$ per τ. Thus, in going from $t = N\tau$ to $t = (N+1)\tau$, the voltage will have moved about 63% of the way from its level at $t = N\tau$ toward its final value. So the voltage across L will have dropped to about 37% after τ, and essentially to zero (0.7%) after about 5τ. Kirchhoff's voltage law implies that the voltage across the resistor will *rise* at the same rate.

When the voltage source is then replaced with a short-circuit, the voltage across R drops exponentially with *t* from *V* towards 0. R will be discharged to about 37% after τ, and essentially fully discharged (0.7%) after about 5τ. Note that the current, *I*, in the circuit behaves as the voltage across R does, via Ohm's Law.

The delay in the rise/fall time of the circuit is in this case caused by the back-EMF from the inductor which, as the current flowing through it tries to change, prevents the current (and hence the voltage across the resistor) from rising or falling much faster than the time-constant of the circuit. Since all wires have some self-inductance and resistance, all circuits have a time constant. As a result, when the power supply is switched on, the current does not instantaneously reach its steady-state value, $\frac{V}{R}$. The rise instead takes several time-constants to complete. If this were not the case, and the current were to reach steady-state immediately, extremely strong inductive electric fields would be generated by the sharp change in the magnetic field — this would lead to breakdown of the air in the circuit and electric arcing, probably damaging components (and users).

These results may also be derived by solving the differential equation describing the circuit:

$$V_{in} = IR + L\frac{dI}{dt}$$

$$V_R = V_{in} - V_L$$

The first equation is solved by using an integrating factor and yields the current which must be differentiated to give V_L; the second equation is straightforward. The solutions are exactly the same as those obtained via Laplace transforms.

Parallel Circuit

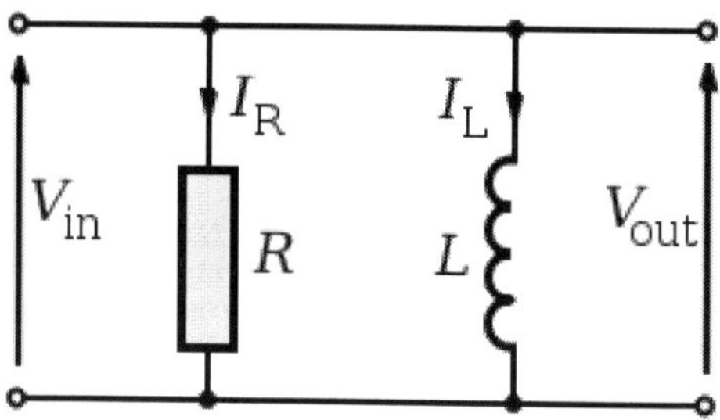

Figure: *Parallel RL circuit*

The parallel RL circuit is generally of less interest than the series circuit unless fed by a current source. This is largely because the output voltage V_{out} is equal to the input voltage V_{in} —as a result, this circuit does not act as a filter for a voltage input signal.

With complex impedances:

$$I_R = \frac{V_{in}}{R}$$

$$I_L = \frac{V_{in}}{j\omega L} = -\frac{jV_{in}}{\omega L}$$

This shows that the inductor lags the resistor (and source) current by 90°.

The parallel circuit is seen on the output of many amplifier circuits, and is used to isolate the amplifier from capacitive loading effects at high frequencies. Because of the phase shift introduced by capacitance, some amplifiers become unstable at very high frequencies, and tend to oscillate. This affects sound quality and component life (especially the transistors), and is to be avoided.

RLC Circuit

An RLC circuit (the letters R, L and C can be in other orders) is an electrical circuit consisting of a resistor, an inductor, and a capacitor, connected in series or in parallel. The RLC part of the name is due to those letters being the usual electrical symbols for resistance, inductance and capacitance respectively.

The circuit forms a harmonic oscillator for current and will resonate in a similar way as an LC circuit will. The main difference that the presence of the resistor makes is that any oscillation induced in the circuit will die away over time if it is not kept going by a source. This effect of the resistor is called damping. The presence of the resistance also reduces the peak resonant frequency somewhat. Some resistance is unavoidable in real circuits, even if a resistor is not specifically included as a component. An ideal, pure LC circuit is an abstraction for the purpose of theory.

There are many applications for this circuit. They are used in many different types of oscillator circuits. Another important application is for tuning, such as in radio receivers or television sets, where they are used to select a narrow range of frequencies from the ambient radio waves. In this role the circuit is often referred to as a tuned circuit. An RLC circuit can be used as a band-pass filter, band-stop filter, low-pass filter or high-pass filter. The tuning application, for instance, is an example of band-pass filtering. The RLC filter is described as a *second-order* circuit, meaning that any voltage or current in the circuit can be described by a second-order differential equation in circuit analysis.

The three circuit elements can be combined in a number of different topologies. All three elements in series or all three elements in parallel are the simplest in concept and the most straightforward to analyse. There are, however, other arrangements, some with practical importance in real circuits. One issue often encountered is the need to take into account inductor resistance. Inductors are typically constructed from coils of wire, the resistance of which is not usually desirable, but it often has a significant effect on the circuit.

Basic Concepts

Resonance

An important property of this circuit is its ability to resonate at a specific frequency, the resonance frequency, f_0 . Frequencies are measured in units of hertz. In this article, however, angular frequency, ω_0 , is used which is more mathematically convenient. This is measured in radians per second. They are related to each other by a simple proportion,

$$\omega_0 = 2\pi f_0$$

Resonance occurs because energy is stored in two different ways: in an electric field as the capacitor is charged and in a magnetic field as current flows through the inductor. Energy can be transferred from one to the other within the circuit and this can be oscillatory. A mechanical analogy is a weight suspended on a spring which will oscillate up and down when released. This is no passing metaphor; a weight on a spring is described by exactly the same second order differential equation as an RLC circuit and for all the properties of the one system there will be found an analogous property of the other. The mechanical property answering to the resistor in the circuit is friction in the spring/weight system. Friction will slowly bring any oscillation to a halt if there is no external force driving it. Likewise, the resistance in an RLC circuit will "damp" the oscillation, diminishing it with time if there is no driving AC power source in the circuit.

The resonance frequency is defined as the frequency at which the impedance of the circuit is at a minimum. Equivalently, it can be defined as the frequency at which the impedance is purely real (that is, purely resistive). This occurs because the impedances of the inductor and capacitor at resonance are equal but of opposite sign and cancel out. Circuits where L and C are in parallel rather than series actually have a maximum impedance rather than a minimum impedance. For this reason they are often described as antiresonators, it is still usual, however, to name the frequency at which this occurs as the resonance frequency.

Natural Frequency

The resonance frequency is defined in terms of the impedance presented to a driving source. It is still possible for the circuit to carry on oscillating (for a time) after the driving source has been removed or it is subjected to a step in voltage (including a step down to zero). This is similar to the way that a tuning fork will carry on ringing after it has been struck, and the effect is often called ringing. This effect is the peak natural resonance frequency of the circuit and in general is not exactly the same as the driven resonance frequency, although the two will usually be quite close to each other. Various terms are used by different authors to distinguish the two, but resonance frequency unqualified usually means the driven resonance frequency. The driven frequency may be called the undamped resonance frequency or undamped natural frequency and the peak frequency may be called the damped resonance frequency or the damped natural frequency. The reason for this terminology is that the driven resonance frequency in a series or parallel resonant circuit has the value

$$\omega_0 = \frac{1}{\sqrt{LC}}$$

This is exactly the same as the resonance frequency of an LC circuit, that is, one with no resistor present. The resonant frequency for an RLC circuit is the same as a circuit in which there is no damping, hence undamped resonance frequency. The peak resonance frequency, on the other hand, depends on the value of the resistor and is described as the damped resonant frequency. A highly damped circuit will fail to resonate at all when not driven. A circuit with a value of resistor that causes it to be just on the edge of ringing is called critically damped. Either side of critically damped are described as underdamped (ringing happens) and overdamped (ringing is suppressed).

Circuits with topologies more complex than straightforward series or parallel (some examples described later in the article) have a driven resonance frequency that deviates from $\omega_0 = \frac{1}{\sqrt{LC}}$ and for those the undamped resonance frequency, damped resonance frequency and driven resonance frequency can all be different.

Damping

Damping is caused by the resistance in the circuit. It determines whether or not the circuit will resonate naturally (that is, without a driving source). Circuits which will resonate in this way are described as underdamped and those that will not are overdamped. Damping attenuation (symbol α) is measured in nepers per second. However, the unitless damping factor (symbol ζ, zeta) is often a more useful measure, which is related to α by

$$\zeta = \frac{\alpha}{\omega_0}$$

The special case of $\zeta = 1$ is called critical damping and represents the case of a circuit that is just on the border of oscillation. It is the minimum damping that can be applied without causing oscillation.

Bandwidth

The resonance effect can be used for filtering, the rapid change in impedance near resonance can be used to pass or block signals close to the resonance frequency. Both band-pass and band-stop filters can be constructed and some filter circuits are shown later in the article. A key parameter in filter design is bandwidth. The bandwidth is measured between the 3dB-points, that is, the frequencies at which the power passed through the circuit has fallen to half the value passed at resonance. There are two of these half-power frequencies, one above, and one below the resonance frequency

$$\Delta\omega = \omega_2 - \omega_1$$

where $\Delta\omega$ is the bandwidth, ω_1 is the lower half-power frequency and ω_2 is the upper half-power frequency. The bandwidth is related to attenuation by,

$$\Delta\omega = 2\alpha$$

when the units are radians per second and nepers per second respectively. Other units may require a conversion factor. A more general measure of bandwidth is the fractional bandwidth, which expresses the bandwidth as a fraction of the resonance frequency and is given by

$$F_b = \frac{\Delta\omega}{\omega_0}$$

The fractional bandwidth is also often stated as a percentage. The damping of filter circuits is adjusted to result in the required bandwidth. A narrow band filter, such as a notch filter, requires low damping. A wide band filter requires high damping.

Q factor

The Q factor is a widespread measure used to characterise resonators. It is defined as the peak energy stored in the circuit divided by the average energy dissipated in it per radian at resonance. Low Q circuits are therefore damped and lossy and high Q circuits are underdamped. Q is related to bandwidth; low Q circuits are wide band and high Q circuits are narrow band. In fact, it happens that Q is the inverse of fractional bandwidth

$$Q = \frac{1}{F_b} = \frac{\omega_0}{\Delta\omega}$$

Q factor is directly proportional to selectivity, as Q factor depends inversely on bandwidth.

For a series resonant circuit, the Q factor can be calculated as follows:

$$Q = \frac{1}{\omega_0 RC} = \frac{\omega_0 L}{R}$$

Scaled Parameters

The parameters ζ, F_b, and Q are all scaled to ω_0. This means that circuits which have similar parameters share similar characteristics regardless of whether or not they are operating in the same frequency band.

The article next gives the analysis for the series RLC circuit in detail. Other configurations are not described in such detail, but the key differences from the series case are given. The general form of the differential equations given in the series circuit section are applicable to all second order circuits and can be used to describe the voltage or current in any element of each circuit.

Chapter 5

Electrical Wiring

Building wiring is the electrical wiring and associated devices such as switches, metres and light fittings used in buildings or other structures. Electrical wiring uses insulated conductors.

Wiring safety codes vary by country, and the International Electrotechnical Commission (IEC) is attempting to standardize wiring amongst member countries. Wires and cables are rated by the circuit voltage, temperature, and environmental conditions (moisture, sunlight, oil, chemicals) in which they can be used. Colour codes are used to distinguish line, neutral and ground (earth) wires.

Wiring Safety Codes

Wiring safety codes are intended to protect people and property from electrical shock and fire hazards. Regulations may be established by city, county, provincial/state or national legislation, usually by adopting a model code (with or without local amendments) produced by a technical standards-setting organization, or by a national standard electrical code. Electrical codes arose in the 1880s with the commercial introduction of electrical power. Many conflicting standards existed for the selection of wire sizes and other design rules for electrical installations.

The first electrical codes in the United States originated in New York in 1881 to regulate installations of electric lighting. Since 1897 the US National Fire Protection Association, a private non-profit association formed by insurance companies, has published the *National Electrical Code* (NEC). States, counties or cities often include the NEC in their local building codes by reference along with local differences. The NEC is modified every three years. It is a consensus code considering suggestions from interested parties. The proposals are studied by committees of engineers, tradesmen, manufacturer representatives, fire fighters, and other invitees.

Since 1927, the Canadian Standards Association (CSA) has produced the Canadian *Safety Standard for Electrical Installations*, which is the basis for provincial electrical codes. The CSA also produces the Canadian Electrical Code, the 2006 edition of which references IEC 60364 (*Electrical Installations for Buildings*) and states that the code addresses the fundamental principles of electrical protection in Section 131. The Canadian code reprints Chapter 13 of IEC 60364, but there are no numerical criteria listed in that chapter to assess the adequacy of any electrical installation.

Although the US and Canadian national standards deal with the same physical phenomena and broadly similar objectives, they differ occasionally in technical detail. As part of the North American Free Trade Agreement (NAFTA) program, US and Canadian standards are slowly converging toward each other, in a process known as harmonization.

In European countries, an attempt has been made to harmonize national wiring standards in an IEC standard, IEC 60364 *Electrical Installations for Buildings*. Hence national standards follow an identical system of sections and chapters. However, this standard is not written in such language that it can readily be adopted as a national wiring code. Neither is it designed for field use by electrical tradesmen and inspectors for testing compliance with national wiring standards. By contrast, national codes, such as the NEC or CSA C22.1, generally exemplify the common objectives of IEC 60364, but provide specific rules in a form that allows for guidance of those installing and inspecting electrical systems.

In Germany, DKE (the German Commission for Electrical, Electronic and Information Technologies of DIN and VDE) is the organisation responsible for the promulgation of electrical standards and safety specifications. DIN VDE 0100 is the German wiring regulations document harmonised with IEC 60364.

In the United Kingdom, wiring installations are regulated by the Institution of Engineering and Technology *Requirements for Electrical Installations: IEE Wiring Regulations, BS 7671: 2008,* which are harmonised with IEC 60364. The 17th edition (issued in January 2008) includes new sections for microgeneration and solar photovoltaic systems. The first edition was published in 1882.

In Australia and New Zealand, the AS/NZS 3000 standard, commonly known as the "wiring rules", specifies requirements for the selection and installation of electrical equipment, and the design and testing of such installations. The standard is mandatory in both New Zealand and Australia; therefore, all electrical work covered by the standard must comply.

The international standard wire sizes are given in the IEC 60228 standard of the International Electrotechnical Commission. In North America, the American Wire Gauge standard for wire sizes is used.

Colour Code

To enable wires to be easily and safely identified, all common wiring safety codes mandate a colour scheme for the insulation on power conductors. In a typical electrical code, some colour-coding is mandatory, while some may be optional. Many local rules and exceptions exist. Older installations vary in colour codes, and colours may shift with insulation exposure to heat, light, and ageing.

Cables

Armoured cables with two rubber-insulated conductors in a flexible metal sheath were used as early as 1906, and were considered at the time a better method than open knob-and-tube wiring, although much more expensive.

The first rubber-insulated cables for building wiring were introduced in 1922 with US patent 1458803, Burley, Harry & Rooney, Henry, "Insulated electric wire", issued 1923-06-12, assigned to Boston Insulated Wire And Cable. These were two or more solid copper electrical wires with rubber insulation, plus woven cotton cloth over each conductor for protection of the insulation, with an overall woven jacket, usually impregnated with tar as a protection from moisture. Waxed paper was used as a filler and separator.

Over time, rubber-insulated cables become brittle because of exposure to atmospheric oxygen, so they must be handled with care, and are usually replaced during renovations. When switches, outlets or light fixtures are replaced, the mere act of tightening connections may cause hardened insulation to flake off the conductors. Rubber insulation further inside the cable often is in better condition than the insulation exposed at connections, due to reduced exposure to oxygen.

Rubber insulation was hard to strip from bare copper, so copper was tinned, causing slightly more electrical resistance. Rubber insulation is no longer used for permanent wiring installations, but may still be used for replaceable temporary cables where flexibility is important, such as electrical extension cords.

About 1950, PVC insulation and jackets were introduced, especially for residential wiring. About the same time, single conductors with a thinner PVC insulation and a thin nylon jacket (e.g. US Type THN, THHN, etc.) became common.

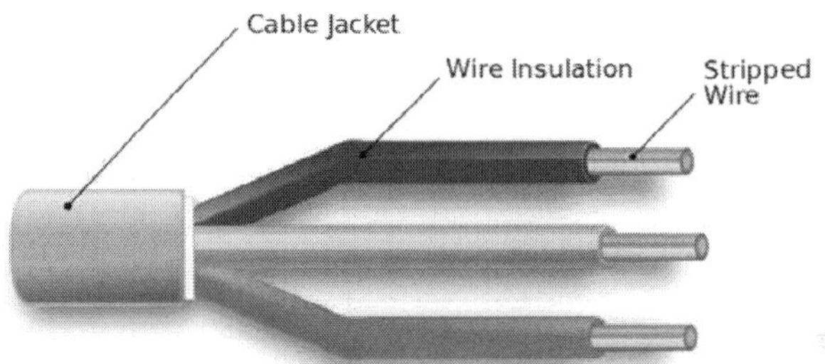

Figure: *Diagram of a simple electrical cable with three insulated conductors*

The simplest form of cable has two insulated conductors twisted together to form a unit; such unjacketed cables with two or three conductors are used for low-voltage signal and control applications such as doorbell wiring. In North American practice, an overhead cable from a transformer on a power pole to a residential electrical service consists of three twisted (triplexed) wires, often with one being a bare wire made of copper (protective earth/ground) and the other two being insulated for the line voltage (hot/line wire and neutral wire). For additional safety, the ground wire may be formed into a stranded co-axial layer completely surrounding the phase conductors, so that the outmost conductor is grounded.

Copper Conductors

Electrical devices often contain copper conductors because of their multiple beneficial properties, including their high electrical conductivity, tensile strength, ductility, creep resistance, corrosion resistance, thermal conductivity, coefficient of thermal expansion, solderability, resistance to electrical overloads, compatibility with electrical insulators, and ease of installation.

Despite competition from other materials, copper remains the preferred electrical conductor in nearly all categories of electrical wiring. For example, copper is used to conduct electricity in high, medium and low voltage power networks, including power generation, power transmission, power distribution, telecommunications, electronics circuitry, data processing, instrumentation, appliances, entertainment systems, motors, transformers, heavy industrial machinery, and countless other types of electrical equipment.

Aluminium Conductors

Aluminium wire was common in North American residential wiring from the late 1960s to mid-1970s due to the rising cost of copper. Because of its greater resistivity, aluminium wiring requires larger conductors than copper. For instance, instead of 14 AWG (American wire gauge) for most lighting circuits, aluminium wiring would be 12 AWG on a typical 15 ampere circuit, though local building codes may vary.

Aluminium conductors were originally indiscriminately used with wiring devices intended for copper conductors. This practice was found to cause defective connections unless the aluminium was one of a special alloy, or all devices — breakers, switches, receptacles, splice connectors, wire nuts, etc. — were specially designed for the purpose. These special designs address problems with junctions between dissimilar metals, oxidation on metal surfaces, and mechanical effects that occur as different metals expand at different rates with increases in temperature. Unlike copper, aluminium has a tendency to cold-flow under pressure, so screw clamped connections may get loose over time. This can be mitigated by using spring-loaded connectors that apply constant pressure, applying high pressure cold joints in splices and termination fittings, and torquing the bolted connection.

Also unlike copper, aluminium forms an insulating oxide layer on the surface. This is sometimes addressed by coating aluminium wires with an antioxidant paste at joints, or by applying a mechanical termination designed to break through the oxide layer during installation. Because of improper design and installation, some junctions to wiring devices would overheat under heavy current load, and cause fires. Revised standards for wiring devices (such as the CO/ALR "copper-aluminium-revised" designation) were developed to reduce these problems. Nonetheless, aluminium wiring for residential use has acquired a poor reputation and has fallen out of favour.

Aluminium conductors are still used for bulk power distribution and large feeder circuits, because they cost less than copper wiring, and weigh less, especially in the large sizes needed for heavy current loads. Aluminium conductors must be installed with compatible connectors.

Modern Wiring Materials

Modern non-metallic sheathed cables, such as (US and Canadian) Types NMB and NMC, consist of two to four wires covered with thermoplastic insulation, plus a bare wire for grounding (bonding), surrounded by a flexible plastic jacket. Some versions wrap the individual conductors in paper before the plastic jacket is applied.

Special versions of non-metallic sheathed cables, such as US Type UF, are designed for direct underground burial (often with separate mechanical protection) or exterior use where exposure to ultraviolet radiation (UV) is a possibility. These cables differ in having a moisture-resistant construction, lacking paper or other absorbent fillers, and being formulated for UV resistance.

Rubber-like synthetic polymer insulation is used in industrial cables and power cables installed underground because of its superior moisture resistance. Insulated cables are rated by their allowable operating voltage and their maximum operating temperature at the conductor surface. A cable may carry multiple usage ratings for applications, for example, one rating for dry installations and another when exposed to moisture or oil.

Generally, single conductor building wire in small sizes is solid wire, since the wiring is not required to be very flexible. Building wire conductors larger than 10 AWG (or about 6 mm^2) are stranded for flexibility during installation, but are not sufficiently pliable to use as appliance cord.

Cables for industrial, commercial, and apartment buildings may contain many insulated conductors in an overall jacket, with helical tape steel or aluminium armour, or steel wire armour, and perhaps as well an overall PVC or lead jacket for protection from moisture and physical damage. Cables intended for very flexible service or in marine applications may be protected by woven bronze wires. Power or communications cables (e.g., computer networking) that are routed in or through air-handling spaces (plenums) of office buildings are required under the model building code to be either encased in metal conduit, or rated for low flame and smoke production.

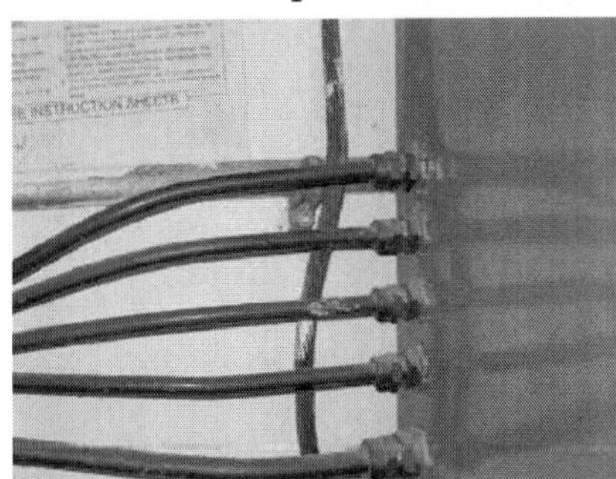

Figure: *Mineral insulated cables at a panel board*

For some industrial uses in steel mills and similar hot environments, no organic material gives satisfactory service. Cables insulated with compressed mica flakes are sometimes used. Another form of high-temperature cable is a mineral insulated cable, with individual conductors placed within a copper tube, and the space filled with magnesium oxide powder. The whole assembly is drawn down to smaller sizes, thereby compressing the powder. Such cables have a certified fire resistance rating, and are more costly than non-fire rated cable. They have little flexibility and behave more like rigid conduit rather than flexible cables.

Because multiple conductors bundled in a cable cannot dissipate heat as easily as single insulated conductors, those circuits are always rated at a lower "ampacity". Tables in electrical safety codes give the maximum allowable current for a particular size of conductor, for the voltage and temperature rating at the surface of the conductor for a given physical environment, including the insulation type and thickness. The allowable current will be different for wet or dry, for hot (attic) or cool (underground) locations. In a run of cable through several areas, the most severe area will determine the appropriate rating of the overall run.

Cables usually are secured by special fittings where they enter electrical apparatus; this may be a simple screw clamp for jacketed cables in a dry location, or a polymer-gasketed cable connector that mechanically engages the armour of an armoured cable and provides a water-resistant connection. Special cable fittings may be applied to prevent explosive gases from flowing in the interior of jacketed cables, where the cable passes through areas where inflammable gases are present. To prevent loosening of the connections of individual conductors of a cable, cables must be supported near their entrance to devices and at regular intervals through their length. In tall buildings, special designs are required to support the conductors of vertical runs of cable. Usually, only one cable per fitting is allowed unless the fitting is otherwise rated.

Special cable constructions and termination techniques are required for cables installed in ocean-going vessels; in addition to electrical safety and fire safety, such cables may also be required to be pressure-resistant where they penetrate bulkheads of a ship. Resistance to corrosion caused by salt water or salt spray is also required.

Raceways

Insulated wires may be run in one of several forms of a raceway between electrical devices. This may be a specialized bendable pipe, called a conduit, or one of several varieties of metal (rigid steel or aluminium) or non-metallic (PVC or HDPE) tubing. Rectangular cross-section metal or PVC wire troughs (North America) or trunking (UK) may be used if many circuits are required. Wires run underground may be run in plastic tubing encased in concrete, but metal elbows may be used in severe pulls. Wiring in exposed areas, for example factory floors, may be run in cable trays or rectangular raceways having lids.

Where wiring, or raceways that hold the wiring, must traverse fire-resistance rated walls and floors, the openings are required by local building codes to be firestopped. In cases where safety-critical wiring must be kept operational during an accidental fire, fireproofing must be applied to maintain circuit integrity in a manner to comply with a product's certification listing. The nature and thickness of any passive fire protection materials used in conjunction with wiring and raceways has a quantifiable impact upon the ampacity derating, because the thermal insulation properties needed for fire resistance also inhibit air cooling of power conductors.

Figure: *A cable tray can be used in stores and dwellings*

Cable trays are used in industrial areas where many insulated cables are run together. Individual cables can exit the tray at any point, simplifying the wiring installation and reducing the labour cost for installing new cables. Power cables may have fittings in the tray to maintain clearance between the conductors, but small control wiring is often installed without any intentional spacing between cables.

Since wires run in conduits or underground cannot dissipate heat as easily as in open air, and since adjacent circuits contribute induced currents, wiring regulations give rules to establish the current capacity (ampacity).

Special sealed fittings are used for wiring routed through potentially explosive atmospheres.

Bus Bars, Bus Duct, Cable Bus

For very heavy currents in electrical apparatus, and for heavy currents distributed through a building, bus bars can be used. (The term "bus" is a contraction of the Latin *omnibus* - meaning "for all".) Each live conductor of such a system is a rigid piece of copper or aluminium, usually in flat bars (but sometimes as tubing or other shapes). Open bus bars are never used in publicly accessible areas, although they are used in manufacturing plants and power company switch yards to gain the benefit of air cooling. A variation is to use heavy cables, especially where it is desirable to transpose or "roll" phases.

In industrial applications, conductor bars are often pre-assembled with insulators in grounded enclosures. This assembly, known as bus duct or busway, can be used for connections to large switchgear or for bringing the main power feed into a building. A form of bus duct known as "plug-in bus" is used to distribute power down the length of a building; it is constructed to allow tap-off switches or motor controllers to be installed at designated places along the bus. The big advantage of this scheme is the ability to remove or add a branch circuit without removing voltage from the whole duct.

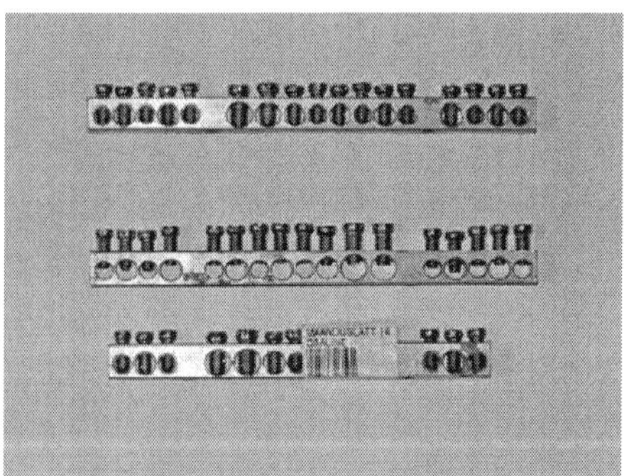

Figure: *Busbars for distributing PE (ground)*

Bus ducts may have all phase conductors in the same enclosure (non-isolated bus), or may have each conductor separated by a grounded barrier from the adjacent phases (segregated bus). For conducting large currents between devices, a cable bus is used.

For very large currents in generating stations or substations, where it is difficult to provide circuit protection, an isolated-phase bus is used. Each phase of the circuit is run in a separate grounded metal enclosure. The only fault possible is a phase-to-ground fault, since the enclosures are separated. This type of bus can be rated up to 50,000 amperes and up to hundreds of kilovolts (during normal service, not just for faults), but is not used for building wiring in the conventional sense.

Electrical Panels

Electrical panels in an electrical service room at St. Mary's Pulp and Paper, Sault Ste. Marie, Ontario, Canada, April 1996

Figure: *Electrical panels, cables and firestops in an electrical service room at a paper mill in Ontario, Canada*

Electrical panels are easily accessible junction boxes used to reroute and switch electrical services. The term is often used to refer to circuit breaker panels or fuseboxes.

Degradation by Pests

Rasberry crazy ants have been known to consume the insides of electrical wiring installations, preferring DC over AC currents. This behaviour is not well understood by scientists.

Squirrels, rats, and other rodents may gnaw on unprotected wiring, causing fire and shock hazards.

Wiring Methods

Materials for wiring interior electrical systems in buildings vary depending on:

- Intended use and amount of power demand on the circuit
- Type of occupancy and size of the building
- National and local regulations
- Environment in which the wiring must operate.

Wiring systems in a single family home or duplex, for example, are simple, with relatively low power requirements, infrequent changes to the building structure and layout, usually with dry, moderate temperature, and non-corrosive environmental conditions. In a light commercial environment, more frequent wiring changes can be expected, large apparatus may be installed, and special conditions of heat or moisture may apply. Heavy industries have more demanding wiring requirements, such as very large currents and higher voltages, frequent changes of equipment layout, corrosive, or wet or explosive atmospheres. In facilities that handle flammable gases or liquids, special rules may govern the installation and wiring of electrical equipment in hazardous areas.

Wires and cables are rated by the circuit voltage, temperature rating, and environmental conditions (moisture, sunlight, oil, chemicals) in which they can be used. A wire or cable has a voltage (to neutral) rating, and a maximum conductor surface temperature rating. The amount of current a cable or wire can safely carry depends on the installation conditions.

Early Wiring Methods

The first interior power wiring systems used conductors that were bare or covered with cloth, which were secured by staples to the framing of the building or on running boards. Where conductors went through walls, they were protected with cloth tape. Splices were done similarly to telegraph connections, and soldered for security. Underground conductors were insulated with wrappings of cloth tape soaked in pitch, and laid in wooden troughs which were then buried. Such wiring systems were unsatisfactory because of the danger of electrocution and fire, plus the high labour cost for such installations.

Other Historical Wiring Methods

Other methods of securing wiring that are now obsolete include:

- Re-use of existing gas pipes when converting gas light installations to electric lighting. Insulated conductors were pulled through the pipes that had formerly supplied the gas lamps. Although used occasionally, this method risked insulation damage from sharp edges inside the pipe at each joint.
- Wood mouldings with grooves cut for single conductor wires, covered by a wooden cap strip. These were prohibited in North American electrical codes by 1928. Wooden moulding was also used to some degree in England, but was never permitted by German and Austrian rules.

- A system of flexible twin cords supported by glass or porcelain buttons was used near the turn of the 20th century in Europe, but was soon replaced by other methods.
- During the first years of the 20th century, various patented forms of wiring system such as Bergman and Peschel tubing were used to protect wiring; these used very thin fibre tubes, or metal tubes which were also used as return conductors.
- In Austria, wires were concealed by embedding a rubber tube in a groove in the wall, plastering over it, then removing the tube and pulling wires through the cavity.

Metal moulding systems, with a flattened oval section consisting of a base strip and a snap-on cap channel, were more costly than open wiring or wooden moulding, but could be easily run on wall surfaces. Similar surface mounted raceway wiring systems are still available today.

Cable Tray

In the electrical wiring of buildings, a cable tray system is used to support insulated electric cables used for power distribution and communication. Cable trays are used as an alternative to open wiring or electrical conduit systems, and are commonly used for cable management in commercial and industrial construction. They are especially useful in situations where changes to a wiring system are anticipated, since new cables can be installed by laying them in the tray, instead of pulling them through a pipe.

Types

Several types of tray are used in different applications. A solid-bottom tray provides the maximum protection to cables, but requires cutting the tray or using fittings to enter or exit cables. A deep, solid enclosure for cables is called a cable channel or cable trough.

A ventilated tray has openings in the bottom of the tray, allowing some air circulation around the cables, water drainage, and allowing some dust to fall through the tray. Small cables may exit the tray through the ventilation openings, which may be either slots or holes punched in the bottom. A ladder tray has the cables supported by a traverse bar, similarly to the rungs of a ladder, at regular intervals on the order of 4 to 12 inches (100 to 300 mm).

Ladder and ventilated trays may have solid covers to protect cables from falling objects, dust, and water. Tray covers for use outdoors or in dusty locations may have a peaked shape to shed debris including dust, ice or snow. Lighter cable trays are more appropriate in situations where a great number of small cables are used, such as for telephone or computer network cables. These trays may be made of wire mesh, called "cable basket", or be designed in the form of a single central spine (rail) with ribs to support the cable on either side.

Large power cables laid in the tray may require support blocks to maintain spacing between conductors, to prevent overheating of the wires. Smaller cables may be laid unsecured in horizontal trays, or secured with cable ties to the bottom of vertically mounted trays.

To maintain support of cables at changes of elevation or direction of a tray, a large number of specialized cable tray fittings are made compatible with each style and manufacturer. Horizontal elbows change direction of a tray in the same plane as the bottom of the tray and are made in 30, 45 and 90 degree forms; inside and outside elbows are for changes perpendicular to the tray bottom. These can be in various shapes including tees and crosses. Some manufacturers and types provide adjustable elbows, useful for field-fitting a tray around obstacles or around irregular shapes.

Various clamping, supporting and splicing accessories are used with the cable tray to provide a complete functional tray system. For example, different sizes of cable tray used within one run can be connected with reducers.

Materials Used

Common cable trays are made of galvanized steel, stainless steel, aluminum, or glass-fibre reinforced plastic. The material for a given application is chosen based on where it will be used. Galvanized tray may be made of pre-galvanized steel sheet fabricated into tray, or may be hot-dip galvanized after fabrication. When galvanized tray is cut to length in the field, usually the cut surface will be painted with a zinc-rich compound to protect the metal from corrosion.

Fire Safety Concerns and Solutions

Combustible cable jackets may catch on fire and cable fires can thus spread along a cable tray within a structure. This is easily prevented through the use of fire-retardant cable jackets, or fireproofing coatings applied to installed cables. Heavy coatings or long fire-stops may require adjustment of the cable current ratings, since such fireproofing measures may reduce the heat dissipation of installed cables.

Regular housecleaning is important for safety, as cable trays are often installed in hard to reach places. Combustible dust and clutter may accumulate if the trays are not routinely checked and kept clean. Plastic and fibreglass reinforced plastic cable trays are combustible; the effect is mitigated through the use of fire retardants or fireproofing.

Ferrous cable trays expand with the increasing heat from accidental fire. This has been proven by the German Otto-Graf-Institut Test Report III.1-80999/Tei/tei "Supplementary Test On The Topic Of Mechanical Force Acting On Cable Penetration Firestop Systems During The Fire Test", dated 23 October 1984, to dislodge "soft" firestops, such as those made of fibrous insulations with rubber coatings. This also applies to any silicone foam seals, but is easily remedied through the use of firestop mortars of sufficient compression strength and thickness, as shown above.

Also, some building codes mandate that penetrants such as cable trays are installed in such ways so as to avoid their contribution to the collapse of a firewall.

Electrical Conduit

An electrical conduit is a tubing system used for protection and routing of electrical wiring. Electrical conduit may be made of metal, plastic, fibre, or fired clay. Flexible conduit is available for special purposes.

Conduit is generally installed by electricians at the site of installation of electrical equipment. Its use, form, and installation details are often specified by wiring regulations, such as the US National Electrical Code (NEC) or other national or local code. The term "conduit" is commonly used by electricians to describe any system that contains electrical conductors, but the term has a more restrictive technical definition when used in official wiring regulations.

History

Some early electric lighting installations made use of existing gas pipe serving gas light fixtures which had been converted to electric lamps. Since this technique provided very good mechanical protection for interior wiring, it was extended to all types of interior wiring and by the early 20th century purpose-built couplings and fittings were manufactured for electrical use.

However, most electrical codes now prohibit the routing of electrical conductors through gas piping, due to concerns about damage to electrical insulation from the rough interiors of pipes and fittings commonly used for gas.

Comparison with other Wiring Methods

Electrical conduit provides very good protection to enclosed conductors from impact, moisture, and chemical vapors. Varying numbers, sizes, and types of conductors can be pulled into a conduit, which simplifies design and construction compared to multiple runs of cables or the expense of customized composite cable. Wiring systems in buildings may be subject to frequent alterations. Frequent wiring changes are made simpler and safer through the use of electrical conduit, as existing conductors can be withdrawn and new conductors installed, with little disruption along the path of the conduit.

A conduit system can be made waterproof or submersible. Metal conduit can be used to shield sensitive circuits from electromagnetic interference, and also can prevent emission of such interference from enclosed power cables. When installed with proper sealing fittings, a conduit will not permit the flow of flammable gases and vapors, which provides protection from fire and explosion hazard in areas handling volatile substances.

Some types of conduit are approved for direct encasement in concrete. This is commonly used in commercial buildings to allow electrical and communication outlets to be installed in the middle of large open areas. For example, retail display cases and open-office areas use floor-mounted conduit boxes to connect power and communications cables. Both metal and plastic conduit can be bent at the job site to allow a neat installation without excessive numbers of manufactured fittings. This is particularly advantageous when following irregular or curved building profiles. Special equipment is used to bend the conduit without kinking or denting it.

The cost of conduit installation is higher than other wiring methods due to the cost of materials and labour. In applications such as residential construction, the high degree of physical damage protection may not be not required, so the expense of conduit is not warranted. Conductors installed within conduit cannot dissipate heat as readily as those installed in open wiring, so the current capacity of each conductor must be reduced (derated) if many are installed in one conduit. It is impractical, and prohibited by wiring regulations, to have more than 360 degrees of total bends in a run of conduit, so special outlet fittings must be provided to allow conductors to be installed without damage in such runs.

Some types of metal conduit may also serve as a useful bonding conductor for grounding (earthing), but wiring regulations may also dictate workmanship standards or supplemental means of grounding for certain types. While metal conduit may sometimes be used as a grounding conductor, the circuit length is limited. For example, a long run of conduit as grounding conductor may have too high an electrical resistance, and not allow proper operation of overcurrent devices on a fault.

Types of Conduit

Conduit systems are classified by the wall thickness, mechanical stiffness, and material used to make the tubing. Materials may be chosen for mechanical protection, corrosion resistance, and overall cost of the installation (labour plus material cost). Wiring regulations for electrical equipment in hazardous areas may require particular types of conduit to be used to provide an approved installation.

Metal Conduits

Rigid Metal Conduit (RMC) is a thick-walled threaded tubing, usually made of coated steel, stainless steel or aluminum.

Galvanized rigid conduit (GRC) is galvanized steel tubing, with a tubing wall that is thick enough to allow it to be threaded. Its common applications are in commercial and industrial construction.

Intermediate Metal Conduit (IMC) is a steel tubing heavier than EMT but lighter than RMC. It may be threaded.

Electrical metallic tubing (EMT), sometimes called thin-wall, is commonly used instead of galvanized rigid conduit (GRC), as it is less costly and lighter than GRC. EMT itself is not threaded, but can be used with threaded fittings that clamp to it. Lengths of conduit are connected to each other and to equipment with clamp-type fittings. Like GRC, EMT is more common in commercial and industrial buildings than in residential applications. EMT is generally made of coated steel, though it may be aluminum.

Aluminum conduit, similar to galvanized steel conduit, is a rigid tube, generally used in commercial and industrial applications where a higher resistance to corrosion is needed. Such locations would include food processing plants, where large amounts of water and cleaning chemicals would make galvanized conduit unsuitable. Aluminum cannot be directly embedded in concrete, since the metal reacts with the alkalis in cement. The conduit may be coated to prevent corrosion by incidental contact with concrete. Aluminum conduit is generally lower cost than steel in addition to having a lower labour cost to install, since a length of aluminum conduit will have about one-third the weight of an equally-sized rigid steel conduit.

In extreme corrosion environments where plastic coating of the tubing is insufficient, conduits may be made from stainless steel, bronze, or brass.

Non-metal Conduits

PVC conduit is the lightest in weight compared to other conduit materials, and usually lower in cost than other forms of conduit. In North American electrical practice, it is available in three different wall thicknesses, with the thin-wall variety only suitable for embedded use in concrete, and heavier grades suitable for direct burial and exposed work. Most of the various fittings made for metal conduit are also available in PVC form. The plastic material resists moisture and many corrosive substances, but since the tubing is non-conductive an extra bonding (grounding) conductor must be pulled into each conduit. PVC conduit may be heated and bent in the field, by using special heating tools designed for the purpose.

Joints to fittings are made with slip-on solvent-welded connections, which set up rapidly after assembly and attain full strength in about one day. Since slip-fit sections do not need to be rotated during assembly, the special union fittings used with threaded conduit (such as Ericson) are not required. Since PVC conduit has a higher coefficient of thermal expansion than other types, it must be mounted to allow for expansion and contraction of each run. Care should be taken when installing PVC underground in multiple or parallel run configurations due to mutual heating effect of densely packed cables, because the conduit will deform when heated.

Rigid Nonmetallic Conduit (RNC) is a non-metallic unthreaded smooth-walled tubing.

Electrical Nonmetallic Tubing (ENT) is a thin-walled corrugated tubing that is moisture-resistant and flame retardant. It is pliable such that it can be bent by hand, and is often flexible although the fittings are not. It is not threaded due to its corrugated shape, although some fittings might be.

Flexible Conduits

Flexible conduits are used to connect to motors or other devices where isolation from vibration is useful, or where an excess number of fittings would be needed to use rigid connections. Electrical codes may restrict the length of a run of some types of flexible conduit.

Flexible Metallic Conduit (FMC, often informally called greenfield or flex) is made by the helical coiling of a self-interlocked ribbed strip of aluminum or steel, forming a hollow tube through which wires can be pulled. FMC is used primarily in dry areas where it would be impractical to install EMT or other non-flexible conduit, yet where metallic strength to protect conductors is still required. The flexible tubing does not maintain any permanent bend, and can flex freely.

FMC may be used as an equipment grounding conductor if specific provisions are met regarding the trade size and length of FMC used, depending on the amperage of the circuits contained in the conduit. In general, an equipment grounding conductor must be pulled through the FMC with an ampacity suitable to carry the fault current likely imposed on the largest circuit contained within the FMC.

Liquidtight Flexible Metal Conduit (LFMC) is a metallic flexible conduit covered by a waterproof plastic coating. The interior is similar to FMC.

Flexible Metallic Tubing (FMT) is not the same as Flexible Metallic Conduit (FMC) which is described in National Electrical Code (NEC) Article 348. FMT is a raceway, but not a conduit and is described in a separate NEC Article 360. It only comes in 1/2" & 3/4" trade sizes, whereas FMC is sized 1/2" ~ 4" trade sizes. NEC 360.2 describes it as: "A raceway that is circular in cross section, flexible, metallic and liquidtight without a nonmetallic jacket."

Liquidtight Flexible Nonmetallic Conduit (LFNC) refers to several types of flame-resistant non-metallic tubing. Interior surfaces may be smooth or corrugated. There may be integral reinforcement within the conduit wall. It is also known as FNMC.

Underground Conduit

Conduit may be installed underground between buildings, structures, or devices to allow installation of power and communication cables. An assembly of these conduits, often called a duct bank, may either be directly buried in earth, or encased in concrete (sometimes with reinforcing rebar to aid against shear forces). Alternatively, a duct bank may be installed in a utility tunnel. A duct bank will allow replacement of damaged cables between buildings or additional power and communications circuits to be added, without the expense of re-excavation of a trench. While metal conduit is occasionally used for burial, usually PVC, polyethylene or polystyrene plastics are now used due to lower cost, easier installation, and better resistance to corrosion.

Formerly, compressed asbestos fibre mixed with cement (such as Transite) was used for some underground installations. Telephone and communications circuits were typically installed in fired-clay conduit.

Cost comparison of some types of conduit

Cost relative to Rigid Galvanized Steel (RGS) conduit, 3/4 inch (21 metric) size

Type	*Labour*	*Weight*	*Material cost*
RMC	1.0	1.0	1.0
Aluminum	0.89	0.34	0.99
IMC	0.89	0.76	0.84
EMT	0.62	0.42	0.35
PVC	0.55	0.20	0.43

Exact ratios of installation labour, weight and material cost vary depending on the size of conduit, but the values for 3/4 inch (21 metric) trade size are representative.

Fittings

Despite the similarity to pipes used in plumbing, purpose-designed electrical fittings are used to connect conduit.

Box connectors join conduit to a junction box or other electrical box. A typical box connector is inserted into a knockout in a junction box, with the threaded end then being secured with a ring (called a lock nut) from within the box, as a bolt would be secured by a nut. The other end of the fitting usually has a screw or compression ring which is tightened down onto the inserted conduit. Fittings for non-threaded conduits are either secured with set screws or with a compression nut that encircles the conduit. Fittings for general purpose use with metal conduits may be made of die-cast zinc, but where stronger fittings are needed, they are made of copper-free aluminum or cast iron.

Couplings connect two pieces of conduit together.

Sometimes the fittings are considered sufficiently conductive to *bond* (electrically unite) the metal conduit to a metal junction box (thus sharing the box's ground connection); other times, grounding bushings are used which have bonding jumpers from the bushing to a grounding screw on the box.

Unlike water piping, if it the conduit is to be watertight, the idea is to keep water *out*, not in. In this case, gaskets are used with special fittings, such as the weatherhead leading from the overhead electrical mains to the electric metre.

Flexible metal conduit usually uses fittings with a clamp on the outside of the box, just like bare cables would.

Conduit Bodies

A conduit body can be used to provide pulling access in a run of conduit, to allow more bends to be made in a particular section of conduit, to conserve space where a full size bend radius would be impractical or impossible, or to split a conduit path into multiple directions. Conductors may *not* be spliced inside a conduit body, unless it is specifically listed for such use.

Conduit bodies differ from junction boxes in that they are not required to be individually supported, this makes them very useful in practical applications. Conduit bodies are commonly referred to as condulets, a term trademarked by Cooper Crouse-Hinds company, a division of Cooper Industries.

Conduit bodies come in various types, moisture ratings, and materials, including galvanized steel, aluminum, and PVC. Depending on the material, they use different mechanical methods for securing conduit. Among the types are:

- L-shaped bodies ("Ells") include the LB, LL, and LR, where the inlet is in line with the access cover and the outlet is on the back, left and right, respectively. In addition to providing access to wires for pulling, "L" fittings allow a 90 degree turn in conduit where there is insufficient space for a full-radius 90 degree sweep (curved conduit section).
- T-shaped bodies ("Tees") feature an inlet in line with the access cover and outlets to both the cover's left and right.
- C-shaped bodies ("Cees") have identical openings above and below the access cover, and are used to pull conductors in a straight runs as they make no turn between inlet and outlet.
- "Service Ell" bodies (SLBs), shorter ells with inlets flush with the access cover, are frequently used where a circuit passes through an exterior wall from outside to inside.

Other Wireways

Surface Mounted Raceway (Wire Molding)

This type of "decorative" conduit is designed to provide an aesthetically acceptable passageway for wiring without hiding it inside or behind a wall. This is used where additional wiring is required, but where going through a wall would be difficult or require remodeling. The conduit has an open face with removable cover, secured to the surface, and wire is placed inside. Plastic raceway is often used for telecommunication wiring, such as network cables in an older structure, where it is not practical to drill through concrete block.

Advantages

- Allows adding new wiring to an existing building without removing or cutting holes into the drywall, lath and plaster, concrete, or other wall finish.
- Allows circuits to be easily locatable and accessible for future changes, thus enabling minimum effort upgrades.

Disadvantages

- Appearance may not be acceptable to all observers.

Trunking

The term *trunking* is used in the United Kingdom for electrical wireways, generally rectangular in cross section with removable lids.

Mini Trunking is a term used in the UK for small form-factor (usually 6 mm to 25 mm square or rectangle sectioned) PVC wireways.

In North American practice "wire trough" or "lay-in wire ways" are terms used to designate similar products. Wall duct raceway is the name give to the type that can be enclosed in a wall.

Innerducts

Innerducts are subducts that can be installed in existing underground conduit systems to provide clean, continuous, low-friction paths for placing optical cables, which have relatively low pulling tension limits. They provide a means for subdividing conventional conduit that was originally designed for single, large-diameter metallic conductor cables into multiple channels for smaller optical cables.

Innerducts are typically small-diameter, semi-flexible subducts. According to Telcordia GR-356, there are three basic types of innerduct: smoothwall, corrugated, and ribbed. These various designs are based on the profile of the inside and outside diameters of the innerduct. The need for a specific characteristic or combination of characteristics, such as pulling strength, flexibility, or the lowest coefficient of friction, dictates the type of innerduct required.

Beyond the basic profiles or contours (smoothwall, corrugated, or ribbed), innerduct is also available in an increasing variety of multiduct designs. Multiduct may be either a composite unit consisting of up to four or six individual innerducts that are held together by some mechanical means, or a single extruded product having multiple channels through which to pull several cables. In either case, the multiduct is coilable, and can be pulled into existing conduit in a manner similar to that of conventional innerduct.

Passive fire Protection

Conduit is of relevance to both firestopping, where they become penetrants, and fireproofing, where circuit integrity measures can be applied on the outside to keep the internal cables operational during an accidental fire. The British standard BS476 also considers internal fires, whereby the fireproofing must protect the surroundings from cable fires. Any external treatments must consider the effect upon ampacity derating due to internal heat buildup.

Knob and Tube Wiring

Knob and tube wiring (sometimes abbreviated K&T) was an early standardized method of electrical wiring in buildings, in common use in North America from about 1880 to the 1930s. It consisted of single-insulated copper conductors run within wall or ceiling cavities, passing through joist and stud drill-holes via protective porcelain insulating tubes, and supported along their length on nailed-down porcelain knob insulators. Where conductors entered a wiring device such as a lamp or switch, or were pulled into a wall, they were protected by flexible cloth insulating sleeving called loom. The first insulation was asphalt-saturated cotton cloth, then rubber became common. Wire splices in such installations were twisted together for good mechanical strength, then soldered and wrapped with rubber insulating tape and friction tape (asphalt saturated cloth), or made inside metal junction boxes.

Knob and tube wiring was eventually displaced from interior wiring systems because of the high cost of installation compared with use of power cables, which combined both power conductors of a circuit in one run (and which later included grounding conductors).

At present, new knob and tube installations are permitted in the US only in a few very specific situations listed in the National Electrical Code, such as certain industrial and agricultural environments.

Elements

Ceramic knobs were cylindrical and generally nailed directly into the wall studs or floor joists. Most had a circular groove running around their circumference, although some were constructed in two pieces with pass-through grooves on each side of the nail in the middle. A leather washer often cushioned the ceramic, to reduce breakage during installation.

By wrapping electrical wires around the knob, and securing them with tie wires, the knob securely and permanently anchored the wire. The knobs separated the wire from potentially combustible framework, facilitated changes in direction, and ensured that wires were not subject to excessive tension. Because the wires were suspended in air, they could dissipate heat well.

Ceramic tubes were inserted into holes bored in wall studs or floor joists, and the wires were directed through them. This kept the wires from coming into contact with the wood framing members and from being compressed by the wood as the house settled. Ceramic tubes were sometimes also used when wires crossed over each other, for protection in case the upper wire were to break and fall on the lower conductor.

Ceramic cleats, which were block-shaped pieces, served a purpose similar to that of the knobs. Not all knob and tube installations utilized cleats.

Ceramic bushings protected each wire entering a metal device box, when such an enclosure was used.

Loom, a woven flexible insulating sleeve, was slipped over insulated wire to provide additional protection whenever a wire passed over or under another wire, when a wire entered a metal device enclosure, and in other situations prescribed by code.

Other ceramic pieces would typically be used as a junction point between the wiring system proper, and the more flexible cloth-clad wiring found in light fixtures or other permanent, hard-wired devices. When a generic power outlet was desired, the wiring could run directly into the junction box through a tube of protective loom and a ceramic bushing.

Wiring devices such as light switches, receptacle outlets, and lamp sockets were either surface-mounted, suspended, or flush-mounted within walls and ceilings. Only in the last case were metal boxes always used to enclose the wiring and device.

Unusual Wiring Layouts

In many older K&T installations, the supply and return wires were routed separately from each other, rather than being located parallel to and near each other. This direct routing method has the advantage of reduced cost by allowing use of the shortest possible lengths of wire, but the major disadvantage that a detailed building wiring diagram is needed for other electricians to understand multiple interwoven circuits, especially if the wiring is not fully visible throughout its length.

By contrast, modern electrical codes now require that all residential wiring connections be made only inside protective enclosures, such as junction boxes, and that all connections must remain accessible for inspection, troubleshooting, repair, or modification.

For example, consider the installation of a light, controlled by two switches in different locations. In modern practice, the supply and return wires both arrive at the first switch, where the supply is connected to one of the two switching wires (hot/red); the return wire is carried inside the same multi-conductor cable to the distant second switch. From there, a supply wire carries power to the light, and a return wire goes back to location of the second switch, and then retraces the path back to the fusebox.

For older K&T installations, the supply and return wires were not necessarily installed as a pair, and did not necessarily have to be near each other at all. From the fusebox, a single supply wire would go to the first switch. From there, it would divide into two switching wires. At the second, remote switch the two wires would come together to be a single supply wire again. Finally this supply wire would be routed from the second switch to the light fixture, and then a single return wire would take the shortest path through the building back to the fusebox. The foregoing is an example of the so-called Carter system wire layout, which is described in more detail in a separate article.

In the case of modern North American split-phase power, the supply wires from two opposite single-phase circuit breakers are used to supply 240VAC for high power devices. In modern circuits, these two wires are grouped together, and for each 240V device a dedicated circuit is installed, usually with the two circuit breakers ganged together to operate simultaneously.

But with old K&T, a past electrician might have reasoned as follows: "I have a 20 amp circuit on one phase over here, and another 20 amp circuit on an opposite phase over there. Rather than run an entirely new set of wires for this one device, I will just tap off the supply wires of those two circuits and bring them together for this one device." This method of wiring does work and it reduces the amount of wire used, but it makes electrical problems very hard to find and fix, with (possibly undocumented) wires going all over the place within the walls and ceilings, to whatever was the most expedient supply or return wire at the time of installation.

Under the US electrical code, Carter system wiring layouts have now been banned, even for permissible new installations of K&T wiring.

However, electricians must be aware of this older system, which is still present in many existing older electrical installations.

Neutral Fusing

Another practice that was common (and even required) in some older K&T designs was the installation of separate fuses in *both* the hot wire and the neutral (return) wire of an electrical circuit. The failure of a neutral fuse would cut off power flow through the affected circuit, but the hot conductor could still remain hot relative to ground, an unexpected and potentially hazardous situation. Because of the presence of a neutral fuse, the neutral conductor could not be relied on to remain near ground potential, and could in fact be at full line potential instead (via transmission of voltage through a switched-on light bulb, for example).

Modern electrical codes generally do not require a neutral fuse. Instead, they explicitly forbid configurations that might break continuity of the neutral conductor, unless all associated hot conductors are also simultaneously disconnected (for example, with ganged or "tied" circuit breakers).

Advantages

When originally installed in the early 1900s, K&T wiring was less expensive than other wiring methods. For several decades, electricians could choose between using K&T wiring, compared to conduit, armored cable, and metal junction boxes. The conduit methods were known to be of better quality, but their cost was significantly higher than that of K&T. In 1909, flexible armored cable cost about twice as much as K&T, and conduit cost about three times the price of K&T. Knob and tube wiring persisted since it allowed owners to wire a building for electricity at lower cost.

Modern wiring methods assume two or more load-carrying conductors will lie very near each other, as for instance in standard NM-2 cable. Since the load-carrying wires are in close proximity, when they heat up, the heating is shared across the wires, limiting the overall current load they can support. Since the load-carrying wires in K&T wiring are widely spaced, the wires are capable of carrying higher currents than the same conductors in close proximity. When installed correctly, the K&T wires are held away from the structural materials by ceramic insulators.

K&T wiring was commonly insulated with cotton cloth and soft rubber, in addition to the porcelain standoffs. Although the actual wire covering may have degraded over the decades, the porcelain standoffs have a nearly unlimited lifespan and will keep any bare wires safely insulated. Today, porcelain standoffs are still commonly used with bare wire electric fencing for livestock, and such porcelain standoffs carry far higher voltage surges without risk of shorting to ground.

In short, K&T wiring which was installed correctly, and not damaged or incorrectly modified since then, is extremely safe when used within the original current-carrying limits.

Disadvantages

Historically, wiring installation requirements were less demanding in the age of knob-and-tube wiring than today. Compared to modern electrical wiring standards, these are the main technical shortcomings of knob-and-tube wiring methods:

- never included a safety grounding conductor
- did not confine switching to the hot conductor (the so-called *Carter system* places electrical loads *across* the common terminals of a three-way switch pair)
- permitted the use of in-line splices in walls without a junction box (however, this downside is offset by the strong nature of the soldered and taped junctions used at the time).
- Over time, the price of electrician labour grew faster than the cost of materials. This removed the price advantage of K&T methods, especially since they required time-consuming skillful soldering of in-line splices and junctions, and careful hand-wrapping of connections in layers of insulating tape.
- Knob-and-tube wiring can be made with high current carrying capacity. However, most existing residential knob and tube installations, dating to before 1940, have fewer branch circuits than is desired today. While these installations were adequate for the electrical loads at the time of installation, modern households use a range and intensity of electrical equipment unforeseen at the time. Household power use increased dramatically following World War II, due to the wide availability of new electrical appliances and devices.
- Modern home buyers often find that existing K&T systems lack the capacity for today's levels of power use. First-generation wiring systems became susceptible to abuse by homeowners who would replace blown fuses with fuses rated for higher current. This overfusing of the circuits subjects wiring to higher levels of current and risks heat damage.

- Knob-and-tube wiring may also be damaged by building renovations. Its cloth and rubber insulation can dry out and turn brittle. It may also be damaged by rodents and careless activities such as hanging objects from wiring running in accessible areas like basements.
- For those concerned about stray magnetic fields, K&T wiring produces a much stronger effect at a given level of current, since the conductors are separated by a greater distance and their fields do not cancel as well as more closely spaced conductors. According to the theory of magnetic fields, two parallel conductors carrying equal currents in opposite directions form a balanced line, partially cancelling each other's magnetic field at a sufficiently large distance from the pair. As a rule of thumb, if two parallel conductors carrying opposite currents are then separated by 10 times the distance, the stray magnetic field will then extend 10 times further than before.
- Currently, the United States National Electrical Code forbids the use of loose, blown-in, or expanding foam insulation over K&T wiring. This is because K&T is designed to let heat dissipate to the surrounding air. As a result, energy efficiency upgrades that involve insulating previously uninsulated walls usually also require replacement of the wiring in affected homes. However, California, Washington, Nebraska, and Oregon have modified the NEC to conditionally allow insulation around K&T. They did not find a single fire that was attributed to K&T, and permit insulation provided the home first passes inspection by an electrician.
- As existing K&T wiring gets older, insurance companies may deny coverage due to a perception of increased risk. Several companies will not write new homeowners policies at all unless all K&T wiring is replaced, or an electrician certifies that the wiring is in good condition. Also, many institutional lenders are unwilling to finance a home with the relatively low-capacity service typical of K&T wiring, unless the electrical service is upgraded. Partial upgrades, where low demand lighting circuits are left intact, may be acceptable to some insurers.

Electrical Installations for Frams

Typical Electrical Distribution System

Electrical energy can be put to many uses and an increasing number of farms will benefit from electrification as the electrical supply network is expanded in the rural areas or generators are installed at farms. Although few farms, in particular small farms, are connected to an electrical supply at present, everyone concerned with design and construction of farm buildings will need to have an appreciation of the general layout and function of electrical installations.

The layout should indicate where outlets, lighting points, switches, motors, heaters and other appliances are to be fitted and the accompanying specifications should describe the chooser wiring system, fixing heights and detail each appliance. Detailed wiring plans and installation designs prepared by a specialist will only be necessary for large and complex buildings, such as plants for processing of agricultural produce.

Electrical Supply

Electricity supply to a farm will normally reach it overhead from a local transformer substation where the voltage has been redued to a three-phase, 415/ 240V supply. Four wires are required for a three-phase supply, one for each of the lines and one common return or neutral. The neutral is connected to earth at the substation. The voltage between any phase wire and the neutral is 240V, while it is 415V between any two phase wires.

If nearly equal loads are connected to each of the phases, the current in the neutral will be kept to a minimum. To achieve this most appliances that consume large amounts of electricity, notably electrical motors and larger heater and air-conditioning units are designed for connection to a three-phase supply. Lighting circuits, socket outlet circuits and appliances of low power rating are served with single-phase supply, but the various circuits are connected to different phases to balance the overall loading. However, sometimes small farms or domestic houses are served with a single-phase, 240V supply. In this case only two wires are required in the supply cable, one live and one neutral. The balancing of loading is effected at the substation, where the lines from several houses are brought together.

The intake point for the main supply to a farm should be at a convenient place that allow for the possible distribution circuits. The intake point must provide for an easily accessible area that is protected from moisture and dust and where the main fuse, the main switch and the metre can be fitted. Circuit fuses and distribution gear may be fitted at this place or in each buildings at the farmstead that is to be served with electricity.

Electricity tariffs are the charges that are passed on to the consumer. The charges commonly consist of two elements, a fixed cost that often depend on the size of the main fuse, and a running cost that depend on the amount of electric energy consumed. The required amp rating for the main fuse will depend on the maximum sum of power required for appliances that are to be connected at any one time and is also influenced by the type of starter used for electrical motors. Usually the motor having the highest power rating will be the determining factor at a farmstead.

Earthing and Bonding

Should a base live wire touch or otherwise become connected to the metalframe work of an appliance, a person touching this would receive an electric shock. A precaution against this is to connect any exposed metal work to an earth wire, that is lead as an extra conductor in the supply cable and connected to an earthing connector, which consist of a number of copper rods driven well into the ground. An earthing wire will thus be carried as a third conductor in single phase supply cable and as a fourth or fifth conductor in a 3-phase supply cable, depending on whether the cable include a neutral wire. The neutral should not be used for earthing. Some appliances are, instead of being earthed, protected by being enclosed in an insulating cover.

Bonding is a low resistance connection between any two point of an earthed system as to prevent any difference of potential that could produce a current and is an additional protection. If, for example, the metal furnishing in a milking parlour is electrically connected to the reinforcement bars of the concrete floor, the cows will be protected from electrical shocks, should for some reason the furnishing become charged by an earth-leaking current, that is not large enough to blow a fuse, since the floor will get the same electrical potential.

Distribution Circuits

Electricity is distributed within buildings in cables, which consist of one or several conductors made of copper or aluminium each separately surrounded by an insulative material, such as plastic, and then enclosed in an outer sheet of plastic or rubber. The size of a cable is given by the cross-section area of its conductors. All cables are assigned a rating in amperes, which is the maximum load the cable can carry without becoming overheated. Large conductors are usually divided into strands to make the cable more flexible.

Surface wiring is normally used in farm buildings. This implies sheeted cables laid on the surface of walls, ceilings, etc. and fixed with clips. Care must be taken that cables are not sharply bent, are protected when passing through a wall and are laid well away from water pipes. Conduit wiring, where the cables are drawn in concealed tubing, is to expensive and complex to be employed in farm buildings.

Lighting circuits are normally carried out in 5A fusing and wiring (1.0mm^2 cables). While a suitable arrangement of one-way and two-way switches will allow lamps to be switched on and off individually or in groups, each such circuit can serve for example ten 100W lamps without danger of overloading. If all ten lamps are on together they have a power requirement of 1000W. Following the relation:

W=V x A where:

W = power

V = voltage

A = current

The lamps would produce a 4.2A current in a 240V circuit, i.e. it leaves a suitable factor of safety to overloading the fuse and wiring.

Socket circuits are normally carried out in 2.5mm^2 wiring and arranged as ring circuits that are supplied from the mains at both ends through 10 to 15A fuses. In domestic installations a socket circuit can carry any number of outlets provided it does not serve a floor area greater than 100m^2. However, when designing socket circuits for farm buildings, such as the workshop, it will be wise to estimate the current produced by all appliances that are expected to be connected at any one time to avoid overloading. Lamp fittings, switches and outlets are available in a range, offering varying degree of protection against dust and moisture penetration. Although more expensive those offering a high level of protection will normally be required in farm buildings as will fittings positioned outdoors.

No socket outlets are permitted in bathrooms and showers and should be avoided in rooms such as clairies and wash rooms, because of the presence of water.

Fixed electrical apparatus that are single phase supplied, such as water heaters, airconditioners and cookers, should have their own circuits with individual fuses.

Three phase electrical motors and apparatus required power supply cables with four or five conductors, including the earthing wire. Each appliance should have its own power supply and the phase lines must be fused individually. Movable 3-phase motors are supplied from special 3-phase power outlets via a rubber sheeted flexible cord that is fixed to the motor at one end and fitted with a 3-phase plug at the other. All flex cords must be protected from damage by for example wheels and should where possible be hung off the ground.

Flex cords must under no circumstances be connected by twisting the conductors together.

Artificial Lighting

In tropical countries with strong natural light even relatively small windows may provide' sufficient indoor lighting. Hence artificial lighting will mainly be required to extend the hours of light.

The two most commonly used artificial light sources, where electrical energy is available, are in candescent bulbs and flourescent tube. Tubes and fittings for tubes are more expensive than bulbs and bulb fittings, but tubes produce three to five times as much light per unit of electric energy, have up to ten times as long life and have a lower heat production.

Hence flourescent light normally is the cheapest despite the higher initial cost. However, in small rooms where the light is switched on and off frequently bulb fittings are usually preferred as the installation cost in this case is more important than the energy cost.

Merary vapour and sodium lamps are often used for outdoor lighting. They have higher efficiency in terms of light produced than flourescent tubes, but their light covers only a limited spectrum and this tend to distort colours.

Various types of fittings are normally available for both bulbs and tubes. While a naked bulb or tube may be sufficient in some circumstances, fittings that protect the lamp from physical damage and moisture penetration will often be required in farm buildings.

From an optical point of view the fitting should obscure the lamp and present a larger surface area of lower brightness to reduce the glare caused by excessive luminance contrast.

This is particularly important if the lamp is positioned where it will be directly viewed. A lighting point must also be positioned so that reflected glare and trouble some shading of a work area is avoided. While light colours on interior surfaces will create a bright room, shades of blue or green produce a feeling of coolness.

The dusty conditions in many farm buildings implies the use of fittings that allow for easy cleaning. Accumulated dust can reduce the flow of light by more than 50%.

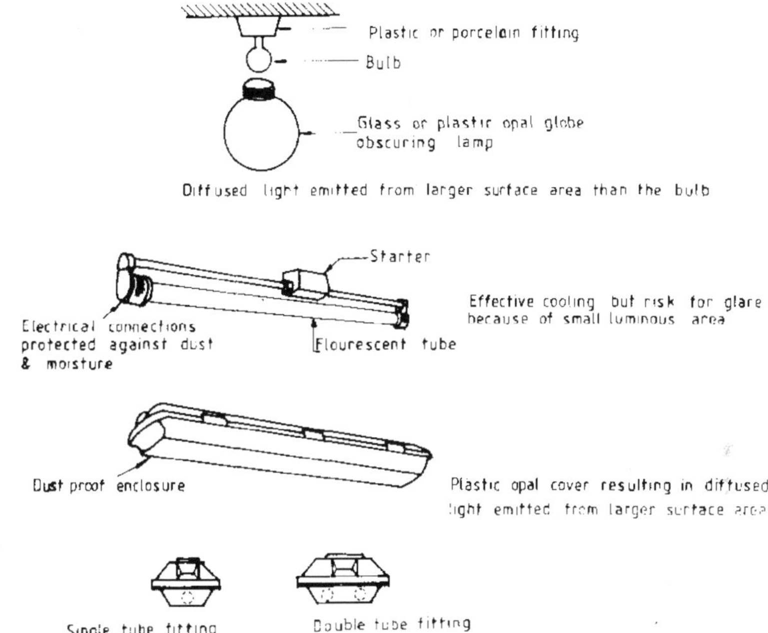

Examples of light Fittings for farm Buildings

Most agricultural production operations carried out in buildings can be performed quite satisfactory using natural light, but where artificial light is to be installed the standard of illumination should be related to the activities carried out. While the installation of 2.0 to 3.0W flourescent light per square metre floor area will be sufficient for general illumination, work areas need more light, say 5 to 8 W/ m^2, and a desk or work bench where concentrated or exacting tasks are performed may need 10 to 15 W/m^2 or more. Where bulbs are to be installed instead of tubes the above values will have to be at least trippled.

Electrical Motors

Single phase motors in sizes up to about 1kW have a wide range of applications, particularly for use in domestic appliances. The most common type, the single phase series motor or universal motor, produce a good starting torque and can be run on both alternating current (AC) and direct current (DC). While it has the advantage of being able to be connected to an ordinary socket outlet, generally, it can not compete with the performance and efficiency of a 3phase motor.

The 3-phase induction motor is the most common electrical motor at farms, where it is used to power fans, transport devises, mills, etc.

Modern electrical motors are manufactured in a wide range of power ratings and types. Their enclosures range from screen protection to totally enclosed. Motors used in farm buildings normally should have an enclosure that is dust-tight and sprinkle-proof, i.e. it should not be damaged by being exposed to sprinkling of water from any direction. However, sometimes even better protection, such as dust-proof and flush-proof, is required and submersible motors must be totally enclosed and completely water proof.

Inherent features of the induction motor are its poor starting torque and heavy starting current - up to six times the full load current. To prevent excessive voltage chop in the supply network and Electricity Company usually allow only small induction motors to be started direct on line. A star/ delta starter is commonly fitted to motor above 2 to 3kW, and will reduce the starting current to about twice the full load current. Unfortunately it also reduces the already poor starting torque even further so that the motor can not start against heavy load. Other types of motors and starters are available for situations where starting against load can not be avoided, however.

The starter for any motor rated above about 0.5kW must incorporate an overload action that switches off should the current exceed a safe value for longer time than what is required to start the motor. In many installations it would, in addition, be desirable to include a release mechanism that prevent unexpected restarting after a power failure. A wide range of sensors, timers and other devises are available for automatic supervision and control of electric motor operation.

Electrical Energy Efficiency on United States Farms

Electrical energy efficiency on United States farms covers the use of electricity on farms and the methods and incentives for improving the efficiency of that use.

U.S. farms have almost doubled their average energy efficiency over the past 25 years.

Usage

Energy costs represent between 2% (at cattle feed lots) and 9% (for grain farming, due partly to grain drying) of farm production costs.

Use on Dairy Farms

In 2006 there were about 65,000 dairy farms in the United States, although most had fewer than 200 cows. One "resource auditor" believes it is possible for dairy farms to reach an energy usage of as low as 200 kWhr per cow per year although an analysis of California dairy farms found that 300 kWhr/year was the lowest actually attained.

One study found the following electrical usages on New York dairy farms:

- Milk cooling – 23%
- Ventilation – 21%
- Vacuum pumps – 18%
- Lighting – 17%
- Electrical water heating – 10%
- Feeding equipment – 7%

Another study found highly similar results.

Regulations

Hitherto almost entirely unregulated (except for zoning ordinances and the effects of property taxes), environmental concerns are now beginning to impose restrictions on farms.

How renewable energy projects may be implemented and also on what rate of return electricity producers can obtain from utilities are key current topics. Acts such as Wisconsin's *Clean Energy Jobs Act* (only proposed as of April, 2010) are controversial among farmers for these reasons. Canada's Green Energy Act 2009 implemented similar kinds of regulations. Contrarily, Carbon offsets may soon offer farmers additional incentives to increase energy use efficiency. However, at least one proposed cap and trade process would raise farmers' costs by up to $1.19 (USD) per acre.

Corporate Average Fuel Economy Fuel standards too (such as in the Energy Independence and Security Act of 2007) are controversial, as biofuels are now a significant portion of farm income.

Efficiency Increases

Increased efficiency in farm usage of electricity is being driven by Greening as well as by economics.

> *"The biggest users of electrical power are heat, light and motors. ... A recent study found that a 3-hp energy efficient compressor used over 42 percent fewer kilowatt-hours (kWhr) than a 3-hp conventional compressor."*

Two prime areas for efficiency improvements are implementing the use of heat recovery systems and Variable Speed Drives. One organization postulated a three-year plan for farms to use energy better.

Crop Rotation

Simply including years of either clover or alfalfa into the cycle of grain crops can yield savings without reducing total crop yields.

Farm Energy Audits

Farms sometimes use accredited "resource auditors" to analyze their operations and recommend improved efficiencies, typically at costs exceeding $1,000 per audit.

However, typical full repayment of an audit's cost could be around six years. Also, a Certified Farm Energy Audit may be required for participation in state or federal energy efficiency programs.

An audit of 20 farms in Cumbria showed savings could be made in all areas examined.

An example of a farm energy audit in Maine is at the following reference. Maryland has a statewide program entitled *EnSave.*

Energy Calculators

Recently a number of "energy calculators" have become available, addressing one or more aspects of farm energy usage. New calculators are being developed under government and utility grants, such as one in Oregon.

Dairy farm Efficiency Improvements

Some improvements are specific to dairy farms, such as using supplemental lighting with energy efficient lamps (which can increase milk production 5 to 16 percent) or relieving heat stress with a combination of sprinklers and fans to increase the cooling effect and improve energy use.

California compiled a "Complete Guide" to dairy farm efficiency improvements.

Cow manure to Electrical Generation

Norswiss Farms of Rice Lake, Wisconsin, an 1100 cow dairy farm, expects to save over $70,000 per year (plus income from electricity sales) from its installation of an 848 kW Combined Heat and Power (CHP) system operating on anaerobic digesters gas from cow manure. The electricity generated is sold to the grid. A similar installation in 2004 on a 270 cow farm in California (one of five in the state at that time) powered a 75 kW generator.

Central Vermont Public Service has a "Cow Power" program using methane-fueled electricity from Vermont farms.

Government and Utility Incentives

U.S. utilities allocated $5.3 billion to energy efficiency programs in 2009.

Government and utility incentives can be beneficial to those entities as well. "It may be less expensive for governments to encourage efficiency and renewable energy options [in order to] delay or negate the need to upgrade and modernize rural grid infrastructure."

A partial list of incentives is available at the following reference: A link to relevant utility pages is at: A third partial list is at:

Alliant Energy in Wisconsin funds the installation of energy-saving projects, receiving payments on the monthly electric bills. National Grid has a similar program.

Electric Motor

An electric motor is an electric machine that converts electrical energy into mechanical energy. The reverse conversion of mechanical energy into electrical energy is done by an electric generator.

In normal motoring mode, most electric motors operate through the interaction between an electric motor's magnetic field and winding currents to generate force within the motor. In certain applications, such as in the transportation industry with traction motors, electric motors can operate in both motoring and generating or braking modes to also produce electrical energy from mechanical energy.

Found in applications as diverse as industrial fans, blowers and pumps, machine tools, household appliances, power tools, and disk drives, electric motors can be powered by direct current (DC) sources, such as from batteries, motor vehicles or rectifiers, or by alternating current (AC) sources, such as from the power grid, inverters or generators. Small motors may be found in electric watches. General-purpose motors with highly standardized dimensions and characteristics provide convenient mechanical power for industrial use. The largest of electric motors are used for ship propulsion, pipeline compression and pumped-storage applications with ratings reaching 100 megawatts. Electric motors may be classified by electric power source type, internal construction, application, type of motion output, and so on.

Electric motors are used to produce linear or rotary force (torque), and should be distinguished from devices such as magnetic solenoids and loudspeakers that convert electricity into motion but do not generate usable mechanical powers, which are respectively referred to as actuators and transducers.

History

Early Motors

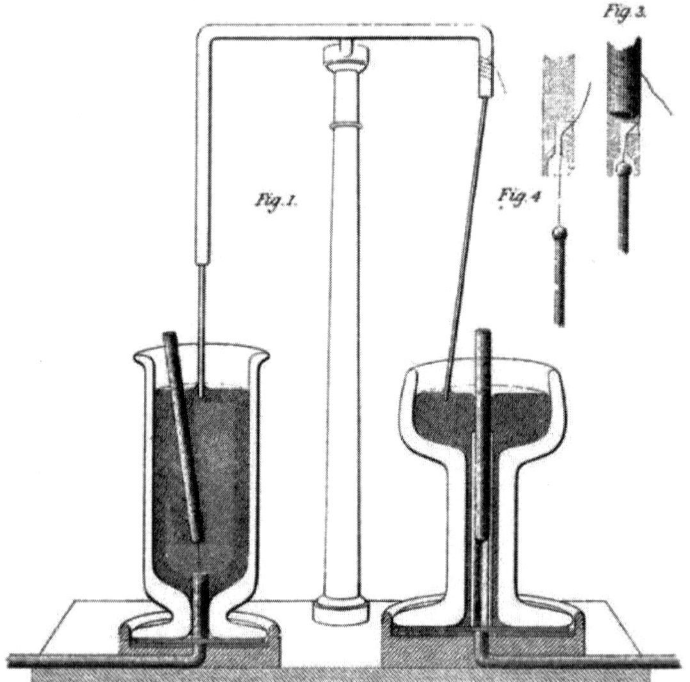

Figure: *Faraday's electromagnetic experiment, 1821*

Perhaps the first electric motors were simple electrostatic devices created by the Scottish monk Andrew Gordon in the 1740s. The theoretical principle behind production of mechanical force by the interactions of an electric current and a magnetic field, Ampère's force law, was discovered later by André-Marie Ampère in 1820. The conversion of electrical energy into mechanical energy by electromagnetic means was demonstrated by the British scientist Michael Faraday in 1821.

A free-hanging wire was dipped into a pool of mercury, on which a permanent magnet (PM) was placed. When a current was passed through the wire, the wire rotated around the magnet, showing that the current gave rise to a close circular magnetic field around the wire. This motor is often demonstrated in physics experiments, brine substituting for toxic mercury. Though Barlow's wheel was an early refinement to this Faraday demonstration, these and similar homopolar motors were to remain unsuited to practical application until late in the century.

In 1827, Hungarian physicist Ányos Jedlik started experimenting with electromagnetic coils. After Jedlik solved the technical problems of the continuous rotation with the invention of commutator, he called his early devices "electromagnetic self-rotors". Although they were used only for instructional purposes, in 1828 Jedlik demonstrated the first device to contain the three main components of practical DC motors: the stator, rotor and commutator. The device employed no permanent magnets, as the magnetic fields of both the stationary and revolving components were produced solely by the currents flowing through their windings.

Success with DC Motors

After many other more or less successful attempts with relatively weak rotating and reciprocating apparatus the German-speaking Prussian Moritz Jacobi created the first real rotating electric motor in May 1834 that actually developed a remarkable mechanical output power. His motor set a world record which was improved only four years later in September 1838 by Jacobi himself. His second motor was powerful enough to drive a boat with 14 people across a wide river. It was not until 1839/40 that other developers worldwide managed to build motors of similar and later also of higher performance.

The first commutator DC electric motor capable of turning machinery was invented by the British scientist William Sturgeon in 1832. Following Sturgeon's work, a commutator-type direct-current electric motor made with the intention of commercial use was built by the American inventor Thomas Davenport, which he patented in 1837. The motors ran at up to 600 revolutions per minute, and powered machine tools and a printing press. Due to the high cost of primary battery power, the motors were commercially unsuccessful and Davenport went bankrupt. Several inventors followed Sturgeon in the development of DC motors but all encountered the same battery power cost issues. No electricity distribution had been developed at the time. Like Sturgeon's motor, there was no practical commercial market for these motors.

In 1855, Jedlik built a device using similar principles to those used in his electromagnetic self-rotors that was capable of useful work. He built a model electric vehicle that same year.

The first commercially successful DC motors followed the invention by Zénobe Gramme who had in 1871 developed the anchor ring dynamo which solved the double-T armature pulsating DC problem. In 1873, Gramme found that this dynamo could be used as a motor, which he demonstrated to great effect at exhibitions in Vienna and Philadelphia by connecting two such DC motors at a distance of up to 2 km away from each other, one as a generator.

In 1886, Frank Julian Sprague invented the first practical DC motor, a non-sparking motor that maintained relatively constant speed under variable loads. Other Sprague electric inventions about this time greatly improved grid electric distribution (prior work done while employed by Thomas Edison), allowed power from electric motors to be returned to the electric grid, provided for electric distribution to trolleys via overhead wires and the trolley pole, and provided controls systems for electric operations.

This allowed Sprague to use electric motors to invent the first electric trolley system in 1887–88 in Richmond VA, the electric elevator and control system in 1892, and the electric subway with independently powered centrally controlled cars, which were first installed in 1892 in Chicago by the South Side Elevated Railway where it became popularly known as the "L". Sprague's motor and related inventions led to an explosion of interest and use in electric motors for industry, while almost simultaneously another great inventor was developing its primary competitor, which would become much more widespread. The development of electric motors of acceptable efficiency was delayed for several decades by failure to recognize the extreme importance of a relatively small air gap between rotor and stator. Efficient designs have a comparatively small air gap. The St. Louis motor, long used in classrooms to illustrate motor principles, is extremely inefficient for the same reason, as well as appearing nothing like a modern motor.

Application of electric motors revolutionized industry. Industrial processes were no longer limited by power transmission using line shafts, belts, compressed air or hydraulic pressure. Instead every machine could be equipped with its own electric motor, providing easy control at the point of use, and improving power transmission efficiency. Electric motors applied in agriculture eliminated human and animal muscle power from such tasks as handling grain or pumping water. Household uses of electric motors reduced heavy labour in the home and made higher standards of convenience, comfort and safety possible. Today, electric motors stand for more than half of the electric energy consumption in the US.

Emergence of AC Motors

In 1824, the French physicist François Arago formulated the existence of rotating magnetic fields, termed Arago's rotations, which, by manually turning switches on and off, Walter Baily demonstrated in 1879 as in effect the first primitive induction motor. In the 1880s, many inventors were trying to develop workable AC motors because AC's advantages in long distance high voltage transmission were counterbalanced by the inability to operate motors on AC. Practical rotating AC induction motors were independently invented by Galileo Ferraris and Nikola Tesla, a working motor model having been demonstrated by the former in 1885 and by the latter in 1887. In 1888, the *Royal Academy of Science of Turin* published Ferraris's research detailing the foundations of motor operation while however concluding that "the apparatus based on that principle could not be of any commercial importance as motor." In 1888, Tesla presented his paper *A New System for Alternating Current Motors and Transformers* to the AIEE that described three patented two-phase four-stator-pole motor types: one with a four-pole rotor forming a non-self-starting reluctance motor, another with a wound rotor forming a self-starting induction motor, and the third a true synchronous motor with separately excited DC supply to rotor winding. One of the patents Tesla filed in 1887, however, also described a shorted-winding-rotor induction motor. George Westinghouse promptly bought Tesla's patents, employed Tesla to develop them, and assigned C. F. Scott to help Tesla, Tesla leaving for other pursuits in 1889.

The constant speed AC induction motor was found not to be suitable for street cars but Westinghouse engineers successfully adapted it to power a mining operation in Telluride, Colorado in 1891. Steadfast in his promotion of three-phase development, Mikhail Dolivo-Dobrovolsky invented the three-phase cage-rotor induction motor in 1889 and the three-limb transformer in 1890. This type of motor is now used for the vast majority of commercial applications. However, he claimed that Tesla's motor was not practical because of two-phase pulsations, which prompted him to persist in his three-phase work. Although Westinghouse achieved its first practical induction motor in 1892 and developed a line of polyphase 60 hertz induction motors in 1893, these early Westinghouse motors were two-phase motors with wound rotors until B. G. Lamme developed a rotating bar winding rotor. The General Electric Company began developing three-phase induction motors in 1891. By 1896, General Electric and Westinghouse signed a cross-licensing agreement for the bar-winding-rotor design, later called the squirrel-cage rotor. Induction motor improvements flowing from these inventions and innovations were such that a 100 horsepower (HP) induction motor currently has the same mounting dimensions as a 7.5 HP motor in 1897.

Motor Construction

Figure: *Electric motor rotor (left) and stator (right)*

Rotor

In an electric motor the moving part is the rotor which turns the shaft to deliver the mechanical power. The rotor usually has conductors laid into it which carry currents that interact with the magnetic field of the stator to generate the forces that turn the shaft. However, some rotors carry permanent magnets, and the stator holds the conductors. Devices such as magnetic solenoids and loudspeakers that convert electricity into motion but do not generate usable mechanical power are respectively referred to as actuators and transducers. Electric motors are used to produce linear force or torque (rotary).

Stator

The stationary part is the stator, usually has either windings or permanent magnets. The stator is the stationary part of the motor's electromagnetic circuit. The stator core is made up of many thin metal sheets, called laminations. Laminations are used to reduce energy losses that would result if a solid core were used.

Air Gap

In between the rotor and stator is the air gap. The air gap has important effects, and is generally as small as possible, as a large gap has a strong negative effect on the performance of an electric motor.

Windings

Windings are wires that are laid in coils, usually wrapped around a laminated soft iron magnetic core so as to form magnetic poles when energized with current.

Electric machines come in two basic magnet field pole configurations: *salient-pole* machine and *nonsalient-pole* machine. In the salient-pole machine the pole's magnetic field is produced by a winding wound around the pole below the pole face. In the *nonsalient-pole*, or distributed field, or round-rotor, machine, the winding is distributed in pole face slots. A shaded-pole motor has a winding around part of the pole that delays the phase of the magnetic field for that pole.

Some motors have conductors which consist of thicker metal, such as bars or sheets of metal, usually copper, although sometimes aluminum is used. These are usually powered by electromagnetic induction.

Commutator

A commutator is a mechanism used to switch the input of certain AC and DC machines consisting of slip ring segments insulated from each other and from the electric motor's shaft. The motor's armature current is supplied through the stationary brushes in contact with the revolving commutator, which causes required current reversal and applies power to the machine in an optimal manner as the rotor rotates from pole to pole.

In absence of such current reversal, the motor would brake to a stop. In light of significant advances in the past few decades due to improved technologies in electronic controller, sensorless control, induction motor, and permanent magnet motor fields, electromechanically commutated motors are increasingly being displaced by externally commutated induction and permanent magnet motors.

Chapter 6

Motor Supply and Control

Motor Supply

A DC motor is usually supplied through slip ring commutator as described above. AC motors' commutation can be either slip ring commutator or externally commutated type, can be fixed-speed or variable-speed control type, and can be synchronous or asynchronous type. Universal motors can run on either AC or DC.

Motor Control

Fixed-speed controlled AC motors are provided with direct-on-line or soft-start starters. Variable speed controlled AC motors are provided with a range of different power inverter, variable-frequency drive or electronic commutator technologies.

The term electronic commutator is usually associated with self-commutated brushless DC motor and switched reluctance motor applications.

Major Categories

Electric motors operate on three different physical principles: magnetic, electrostatic and piezoelectric. By far the most common is magnetic. In magnetic motors, magnetic fields are formed in both the rotor and the stator. The product between these two fields gives rise to a force, and thus a torque on the motor shaft. One, or both, of these fields must be made to change with the rotation of the motor. This is done by switching the poles on and off at the right time, or varying the strength of the pole.

The main types are DC motors and AC motors, the former increasingly being displaced by the latter.

AC electric motors are either asynchronous or synchronous. Once started, a synchronous motor requires synchronism with the moving magnetic field's synchronous speed for all normal torque conditions.

In synchronous machines, the magnetic field must be provided by means other than induction such as from separately excited windings or permanent magnets. It is usual to distinguish motors' rated output power about the unity horsepower threshold so that integral horsepower refers to motor(s) equal to or above, and fractional horsepower (FHP) refers to motor(s) below, the threshold.

Type of Motor Commutation				
Major Categories by Self-Commutated			**Externally Commutated**	
Mechanical-Commutator Motors		**Electronic-Commutator (EC) Motors[59][b]**	**Asynchronous Machines**	**Synchronous Machines[2]**
AC[60][c]	**DC**	**AC[5, 6]**	**AC[6]**	
*** Universal motor (AC commutator series motor[58] or AC/DC motor[57])[1] * Repulsion motor**	**Electrically excited DC motor: * Separately excited * Series * Shunt * Compound PM DC motor**	**With PM rotor: * BLDC motor With ferromagnetic rotor: * SRM**	**Three-phase motors: * SCIM[3, 8] * WRIM[4, 7, 8] AC motors:[10] * Capacitor * Resistance * Split * Shaded-pole**	**Three-phase motors: * WRSM * PMSM or BLAC motor[59] - IPMSM - SPMSM * Hybrid AC motors:[10] * Permanent-split capacitor * Hysteresis * Stepper * SyRM * SyRM-PM hybrid**
Simple electronics	**Rectifier, linear transistor(s) or DC chopper**	**More elaborate electronics**	**Most elaborate electronics (VFD), when provided**	

Notes:

1. Rotation is independent of the frequency of the AC voltage.
2. Rotation is equal to synchronous speed (motor stator field speed).
3. In SCIM fixed-speed operation rotation is equal to slip speed (synchronous speed less slip).
4. In non-slip energy recovery systems WRIM is usually used for motor starting but can be used to vary load speed.
5. Variable-speed operation.
6. Whereas induction and synchronous motor drives are typically with either six-step or sinusoidal waveform output, BLDC motor drives are usually with trapezoidal current waveform; the behaviour of both sinusoidal and trapezoidal PM machines is however identical in terms of their fundamental aspects.
7. In variable-speed operation WRIM is used in slip energy recovery and double-fed induction machine applications.
8. Cage winding refers to shorted-circuited squirrel-cage rotor, wound winding being connected externally through slip rings.
9. Mostly single-phase with some three-phase.

Abbreviations:

- BLAC - Brushless AC
- BLDC - Brushless DC
- BLDM - Brushless DC motor
- EC - Electronic commutator
- PM - Permanent magnet
- IPMSM - Interior permanent magnet synchronous motor
- PMSM - Permanent magnet synchronous motor
- SPMSM - Surface permanent magnet synchronous motor
- SCIM - Squirrel-cage induction motor
- SRM - Switched reluctance motor
- SyRM - Synchronous reluctance motor
- VFD - Variable-frequency drive
- WRIM - Wound-rotor induction motor
- WRSM - Wound-rotor synchronous motor

Self-Commutated Motor

Brushed DC Motor

All self-commutated DC motors are by definition run on DC electric power. Most DC motors are small PM types. They contain a brushed internal mechanical commutation to reverse motor windings' current in synchronism with rotation.

Electrically Excited DC Motor

A commutated DC motor has a set of rotating windings wound on an armature mounted on a rotating shaft. The shaft also carries the commutator, a long-lasting rotary electrical switch that periodically reverses the flow of current in the rotor windings as the shaft rotates. Thus, every brushed DC motor has AC flowing through its rotating windings. Current flows through one or more pairs of brushes that bear on the commutator; the brushes connect an external source of electric power to the rotating armature.

The rotating armature consists of one or more coils of wire wound around a laminated, magnetically "soft" ferromagnetic core. Current from the brushes flows through the commutator and one winding of the armature, making it a temporary magnet (an electromagnet). The magnets field produced by the armature interacts with a stationary magnetic field produced by either PMs or another winding a field coil, as part of the motor frame. The force between the two magnetic fields tends to rotate the motor shaft. The commutator switches power to the coils as the rotor turns, keeping the magnetic poles of the rotor from ever fully aligning with the magnetic poles of the stator field, so that the rotor never stops (like a compass needle does), but rather keeps rotating as long as power is applied.

Many of the limitations of the classic commutator DC motor are due to the need for brushes to press against the commutator. This creates friction. Sparks are created by the brushes making and breaking circuits through the rotor coils as the brushes cross the insulating gaps between commutator sections. Depending on the commutator design, this may include the brushes shorting together adjacent sections – and hence coil ends – momentarily while crossing the gaps.

Furthermore, the inductance of the rotor coils causes the voltage across each to rise when its circuit is opened, increasing the sparking of the brushes. This sparking limits the maximum speed of the machine, as too-rapid sparking will overheat, erode, or even melt the commutator. The current density per unit area of the brushes, in combination with their resistivity, limits the output of the motor. The making and breaking of electric contact also generates electrical noise; sparking generates RFI. Brushes eventually wear out and require replacement, and the commutator itself is subject to wear and maintenance (on larger motors) or replacement (on small motors). The commutator assembly on a large motor is a costly element, requiring precision assembly of many parts. On small motors, the commutator is usually permanently integrated into the rotor, so replacing it usually requires replacing the whole rotor.

While most commutators are cylindrical, some are flat discs consisting of several segments (typically, at least three) mounted on an insulator.

Large brushes are desired for a larger brush contact area to maximize motor output, but small brushes are desired for low mass to maximize the speed at which the motor can run without the brushes excessively bouncing and sparking. (Small brushes are also desirable for lower cost.) Stiffer brush springs can also be used to make brushes of a given mass work at a higher speed, but at the cost of greater friction losses (lower efficiency) and accelerated brush and commutator wear. Therefore, DC motor brush design entails a trade-off between output power, speed, and efficiency/wear.

DC machines are defined as follows:

- Armature circuit - A winding where the load current is carried, such that can be either stationary or rotating part of motor or generator.
- Field circuit - A set of windings that produces a magnetic field so that the electromagnetic induction can take place in electric machines.
- Commutation: A mechanical technique in which rectification can be achieved, or from which DC can be derived, in DC machines.

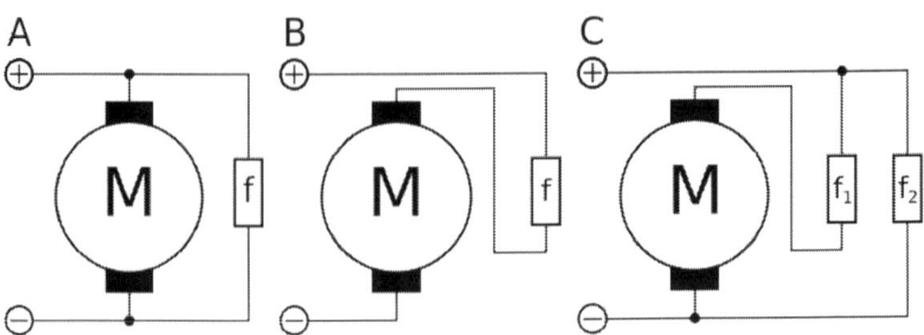

Figure: *A: shunt B: series C: compound f = field coil*

There are five types of brushed DC motor:

- DC shunt-wound motor
- DC series-wound motor
- DC compound motor (two configurations):
  - o Cumulative compound
  - o Differentially compounded
- PM DC motor (not shown)
- Separately excited (not shown).

Permanent Magnet DC Motor

A PM motor does not have a field winding on the stator frame, instead relying on PMs to provide the magnetic field against which the rotor field interacts to produce torque. Compensating windings in series with the armature may be used on large motors to improve commutation under load. Because this field is fixed, it cannot be adjusted for speed control. PM fields (stators) are convenient in miniature motors to eliminate the power consumption of the field winding.

Most larger DC motors are of the "dynamo" type, which have stator windings. Historically, PMs could not be made to retain high flux if they were disassembled; field windings were more practical to obtain the needed amount of flux. However, large PMs are costly, as well as dangerous and difficult to assemble; this favours wound fields for large machines.

To minimize overall weight and size, miniature PM motors may use high energy magnets made with neodymium or other strategic elements; most such are neodymium-iron-boron alloy. With their higher flux density, electric machines with high-energy PMs are at least competitive with all optimally designed singly fed synchronous and induction electric machines. Miniature motors resemble the structure in the illustration, except that they have at least three rotor poles (to ensure starting, regardless of rotor position) and their outer housing is a steel tube that magnetically links the exteriors of the curved field magnets.

Electronic Commutator (EC) Motor

Brushless DC Motor

Some of the problems of the brushed DC motor are eliminated in the BLDC design. In this motor, the mechanical "rotating switch" or commutator is replaced by an external electronic switch synchronised to the rotor's position. BLDC motors are typically 85–90% efficient or more. Efficiency for a BLDC motor of up to 96.5% have been reported, whereas DC motors with brushgear are typically 75–80% efficient.

The BLDC motor's characteristic trapezoidal back-emf waveform is derived partly from the stator windings being evenly distributed, and partly from the placement of the rotor's PMs.

Also known as electronically commutated DC or inside out DC motors, the stator windings of trapezoidal BLDC motors can be with single-phase, two-phase or three-phase and use Hall effect sensors mounted on their windings for rotor position sensing and low cost closed-loop control of the electronic commutator.

BLDC motors are commonly used where precise speed control is necessary, as in computer disk drives or in video cassette recorders, the spindles within CD, CD-ROM (etc.) drives, and mechanisms within office products such as fans, laser printers and photocopiers.

They have several advantages over conventional motors:

- Compared to AC fans using shaded-pole motors, they are very efficient, running much cooler than the equivalent AC motors. This cool operation leads to much-improved life of the fan's bearings.
- Without a commutator to wear out, the life of a BLDC motor can be significantly longer compared to a DC motor using brushes and a commutator. Commutation also tends to cause a great deal of electrical and RF noise; without a commutator or brushes, a BLDC motor may be used in electrically sensitive devices like audio equipment or computers.
- The same Hall effect sensors that provide the commutation can also provide a convenient tachometre signal for closed-loop control (servo-controlled) applications. In fans, the tachometre signal can be used to derive a "fan OK" signal as well as provide running speed feedback.
- The motor can be easily synchronized to an internal or external clock, leading to precise speed control.
- BLDC motors have no chance of sparking, unlike brushed motors, making them better suited to environments with volatile chemicals and fuels. Also, sparking generates ozone which can accumulate in poorly ventilated buildings risking harm to occupants' health.

- BLDC motors are usually used in small equipment such as computers and are generally used in fans to get rid of unwanted heat.
- They are also acoustically very quiet motors which is an advantage if being used in equipment that is affected by vibrations.

Modern BLDC motors range in power from a fraction of a watt to many kilowatts. Larger BLDC motors up to about 100 kW rating are used in electric vehicles. They also find significant use in high-performance electric model aircraft.

Switched Reluctance Motor

The SRM has no brushes or PMs, and the rotor has no electric currents. Instead, torque comes from a slight misalignment of poles on the rotor with poles on the stator. The rotor aligns itself with the magnetic field of the stator, while the stator field stator windings are sequentially energized to rotate the stator field.

The magnetic flux created by the field windings follows the path of least magnetic reluctance, meaning the flux will flow through poles of the rotor that are closest to the energized poles of the stator, thereby magnetizing those poles of the rotor and creating torque. As the rotor turns, different windings will be energized, keeping the rotor turning.

SRMs are now being used in some appliances.

Universal AC-DC Motor

A commutated electrically excited series or parallel wound motor is referred to as a universal motor because it can be designed to operate on both AC and DC power. A universal motor can operate well on AC because the current in both the field and the armature coils (and hence the resultant magnetic fields) will alternate (reverse polarity) in synchronism, and hence the resulting mechanical force will occur in a constant direction of rotation.

Operating at normal power line frequencies, universal motors are often found in a range less than 1000 watts. Universal motors also formed the basis of the traditional railway traction motor in electric railways. In this application, the use of AC to power a motor originally designed to run on DC would lead to efficiency losses due to eddy current heating of their magnetic components, particularly the motor field pole-pieces that, for DC, would have used solid (un-laminated) iron and they are now rarely used.

An advantage of the universal motor is that AC supplies may be used on motors which have some characteristics more common in DC motors, specifically high starting torque and very compact design if high running speeds are used. The negative aspect is the maintenance and short life problems caused by the commutator.

Such motors are used in devices such as food mixers and power tools which are used only intermittently, and often have high starting-torque demands. Multiple taps on the field coil provide (imprecise) stepped speed control.

Household blenders that advertise many speeds frequently combine a field coil with several taps and a diode that can be inserted in series with the motor (causing the motor to run on half-wave rectified AC). Universal motors also lend themselves to electronic speed control and, as such, are an ideal choice for devices like domestic washing machines. The motor can be used to agitate the drum (both forwards and in reverse) by switching the field winding with respect to the armature.

Whereas SCIMs cannot turn a shaft faster than allowed by the power line frequency, universal motors can run at much higher speeds. This makes them useful for appliances such as blenders, vacuum cleaners, and hair dryers where high speed and light weight are desirable. They are also commonly used in portable power tools, such as drills, sanders, circular and jig saws, where the motor's characteristics work well. Many vacuum cleaner and weed trimmer motors exceed 10,000 rpm, while many similar miniature grinders exceed 30,000 rpm.

Externally Commutated AC Machine

The design of AC induction and synchronous motors is optimized for operation on single-phase or polyphase sinusoidal or quasi-sinusoidal waveform power such as supplied for fixed-speed application from the AC power grid or for variable-speed application from VFD controllers. An AC motor has two parts: a stationary stator having coils supplied with AC to produce a rotating magnetic field, and a rotor attached to the output shaft that is given a torque by the rotating field.

Induction Motor

Cage and Wound Rotor Induction Motor

An induction motor is an asynchronous AC motor where power is transferred to the rotor by electromagnetic induction, much like transformer action. An induction motor resembles a rotating transformer, because the stator (stationary part) is essentially the primary side of the transformer and the rotor (rotating part) is the secondary side. Polyphase induction motors are widely used in industry.

Induction motors may be further divided into Squirrel Cage Induction Motors and Wound Rotor Induction Motors. SCIMs have a heavy winding made up of solid bars, usually aluminum or copper, joined by rings at the ends of the rotor. When one considers only the bars and rings as a whole, they are much like an animal's rotating exercise cage, hence the name.

Currents induced into this winding provide the rotor magnetic field. The shape of the rotor bars determines the speed-torque characteristics. At low speeds, the current induced in the squirrel cage is nearly at line frequency and tends to be in the outer parts of the rotor cage. As the motor accelerates, the slip frequency becomes lower, and more current is in the interior of the winding. By shaping the bars to change the resistance of the winding portions in the interior and outer parts of the cage, effectively a variable resistance is inserted in the rotor circuit. However, the majority of such motors have uniform bars.

In a WRIM, the rotor winding is made of many turns of insulated wire and is connected to slip rings on the motor shaft. An external resistor or other control devices can be connected in the rotor circuit. Resistors allow control of the motor speed, although significant power is dissipated in the external resistance. A converter can be fed from the rotor circuit and return the slip-frequency power that would otherwise be wasted back into the power system through an inverter or separate motor-generator.

The WRIM is used primarily to start a high inertia load or a load that requires a very high starting torque across the full speed range. By correctly selecting the resistors used in the secondary resistance or slip ring starter, the motor is able to produce maximum torque at a relatively low supply current from zero speed to full speed. This type of motor also offers controllable speed.

Motor speed can be changed because the torque curve of the motor is effectively modified by the amount of resistance connected to the rotor circuit. Increasing the value of resistance will move the speed of maximum torque down. If the resistance connected to the rotor is increased beyond the point where the maximum torque occurs at zero speed, the torque will be further reduced.

When used with a load that has a torque curve that increases with speed, the motor will operate at the speed where the torque developed by the motor is equal to the load torque. Reducing the load will cause the motor to speed up, and increasing the load will cause the motor to slow down until the load and motor torque are equal. Operated in this manner, the slip losses are dissipated in the secondary resistors and can be very significant. The speed regulation and net efficiency is also very poor.

Torque Motor

A torque motor is a specialised form of electric motor which can operate indefinitely while stalled, that is, with the rotor blocked from turning, without incurring damage. In this mode of operation, the motor will apply a steady torque to the load (hence the name).

A common application of a torque motor would be the supply- and take-up reel motors in a tape drive. In this application, driven from a low voltage, the characteristics of these motors allow a relatively constant light tension to be applied to the tape whether or not the capstan is feeding tape past the tape heads. Driven from a higher voltage, (and so delivering a higher torque), the torque motors can also achieve fast-forward and rewind operation without requiring any additional mechanics such as gears or clutches. In the computer gaming world, torque motors are used in force feedback steering wheels.

Another common application is the control of the throttle of an internal combustion engine in conjunction with an electronic governor. In this usage, the motor works against a return spring to move the throttle in accordance with the output of the governor. The latter monitors engine speed by counting electrical pulses from the ignition system or from a magnetic pickup and, depending on the speed, makes small adjustments to the amount of current applied to the motor. If the engine starts to slow down relative to the desired speed, the current will be increased, the motor will develop more torque, pulling against the return spring and opening the throttle. Should the engine run too fast, the governor will reduce the current being applied to the motor, causing the return spring to pull back and close the throttle.

Synchronous Motor

A synchronous electric motor is an AC motor distinguished by a rotor spinning with coils passing magnets at the same rate as the AC and resulting magnetic field which drives it. Another way of saying this is that it has zero slip under usual operating conditions. Contrast this with an induction motor, which must slip to produce torque. One type of synchronous motor is like an induction motor except the rotor is excited by a DC field. Slip rings and brushes are used to conduct current to the rotor. The rotor poles connect to each other and move at the same speed hence the name synchronous motor. Another type, for low load torque, has flats ground onto a conventional squirrel-cage rotor to create discrete poles. Yet another, such as made by Hammond for its pre-World War II clocks, and in the older Hammond organs, has no rotor windings and discrete poles. It is not self-starting. The clock requires manual starting by a small knob on the back, while the older Hammond organs had an auxiliary starting motor connected by a spring-loaded manually operated switch.

Finally, hysteresis synchronous motors typically are (essentially) two-phase motors with a phase-shifting capacitor for one phase. They start like induction motors, but when slip rate decreases sufficiently, the rotor (a smooth cylinder) becomes temporarily magnetized. Its distributed poles make it act like a PMSM. The rotor material, like that of a common nail, will stay magnetized, but can also be demagnetized with little difficulty. Once running, the rotor poles stay in place; they do not drift.

Low-power synchronous timing motors (such as those for traditional electric clocks) may have multi-pole PM external cup rotors, and use shading coils to provide starting torque. *Telechron* clock motors have shaded poles for starting torque, and a two-spoke ring rotor that performs like a discrete two-pole rotor.

Doubly Fed Electric Machine

Doubly fed electric motors have two independent multiphase winding sets, which contribute active (i.e., working) power to the energy conversion process, with at least one of the winding sets electronically controlled for variable speed operation. Two independent multiphase winding sets (i.e., dual armature) are the maximum provided in a single package without topology duplication. Doubly fed electric motors are machines with an effective constant torque speed range that is twice synchronous speed for a given frequency of excitation. This is twice the constant torque speed range as singly fed electric machines, which have only one active winding set.

A doubly fed motor allows for a smaller electronic converter but the cost of the rotor winding and slip rings may offset the saving in the power electronics components. Difficulties with controlling speed near synchronous speed limit applications.

Special Magnetic Motors

Rotary

Hydraulic Cylinder Displacement: Electric motors are replacing hydraulic cylinders in airplanes and military equipment.

Ironless or Coreless Rotor Motor

Nothing in the principle of any of the motors described above requires that the iron (steel) portions of the rotor actually rotate. If the soft magnetic material of the rotor is made in the form of a cylinder, then (except for the effect of hysteresis) torque is exerted only on the windings of the electromagnets. Taking advantage of this fact is the *coreless or ironless DC motor*, a specialised form of a PM DC motor. Optimized for rapid acceleration, these motors have a rotor that is constructed without any iron core.

The rotor can take the form of a winding-filled cylinder, or a self-supporting structure comprising only the magnet wire and the bonding material. The rotor can fit inside the stator magnets; a magnetically soft stationary cylinder inside the rotor provides a return path for the stator magnetic flux. A second arrangement has the rotor winding basket surrounding the stator magnets. In that design, the rotor fits inside a magnetically soft cylinder that can serve as the housing for the motor, and likewise provides a return path for the flux.

Because the rotor is much lighter in weight (mass) than a conventional rotor formed from copper windings on steel laminations, the rotor can accelerate much more rapidly, often achieving a mechanical time constant under one ms. This is especially true if the windings use aluminum rather than the heavier copper. But because there is no metal mass in the rotor to act as a heat sink, even small coreless motors must often be cooled by forced air. Overheating might be an issue for coreless DC motor designs.

Among these types are the disc-rotor types, described in more detail in the next section.

Vibrator motors for cellular phones are sometimes tiny cylindrical PM field types, but there are also disc-shaped types which have a thin multipolar disc field magnet, and an intentionally unbalanced molded-plastic rotor structure with two bonded coreless coils. Metal brushes and a flat commutator switch power to the rotor coils.

Related limited-travel actuators have no core and a bonded coil placed between the poles of high-flux thin PMs. These are the fast head positioners for rigid-disk ("hard disk") drives. Although the contemporary design differs considerably from that of loudspeakers, it is still loosely (and incorrectly) referred to as a "voice coil" structure, because some earlier rigid-disk-drive heads moved in straight lines, and had a drive structure much like that of a loudspeaker.

Pancake or Axial Rotor Motor

A rather unusual motor design, the printed armature or pancake motor has the windings shaped as a disc running between arrays of high-flux magnets.

The magnets are arranged in a circle facing the rotor with space in between to form an axial air gap. This design is commonly known as the pancake motor because of its extremely flat profile, although the technology has had many brand names since its inception, such as ServoDisc.

The printed armature (originally formed on a printed circuit board) in a printed armature motor is made from punched copper sheets that are laminated together using advanced composites to form a thin rigid disc. The printed armature has a unique construction in the brushed motor world in that it does not have a separate ring commutator. The brushes run directly on the armature surface making the whole design very compact.

An alternative manufacturing method is to use wound copper wire laid flat with a central conventional commutator, in a flower and petal shape. The windings are typically stabilized by being impregnated with electrical epoxy potting systems.

These are filled epoxies that have moderate mixed viscosity and a long gel time. They are highlighted by low shrinkage and low exotherm, and are typically UL 1446 recognised as a potting compound insulated with 180 °C, Class H rating.

The unique advantage of ironless DC motors is that there is no cogging (torque variations caused by changing attraction between the iron and the magnets). Parasitic eddy currents cannot form in the rotor as it is totally ironless, although iron rotors are laminated. This can greatly improve efficiency, but variable-speed controllers must use a higher switching rate (>40 kHz) or DC because of the decreased electromagnetic induction.

These motors were originally invented to drive the capstan(s) of magnetic tape drives in the burgeoning computer industry, where minimal time to reach operating speed and minimal stopping distance were critical. Pancake motors are still widely used in high-performance servo-controlled systems, robotic systems, industrial automation and medical devices.

Due to the variety of constructions now available, the technology is used in applications from high temperature military to low cost pump and basic servos.

Servo Motor

A servomotor is a motor, very often sold as a complete module, which is used within a position-control or speed-control feedback control system mainly control valves, such as motor operated control valves. Servomotors are used in applications such as machine tools, pen plotters, and other process systems.

Motors intended for use in a servomechanism must have well-documented characteristics for speed, torque, and power.

The speed vs. torque curve is quite important and is high ratio for a servo motor. Dynamic response characteristics such as winding inductance and rotor inertia are also important; these factors limit the overall performance of the servomechanism loop.

Large, powerful, but slow-responding servo loops may use conventional AC or DC motors and drive systems with position or speed feedback on the motor. As dynamic response requirements increase, more specialised motor designs such as coreless motors are used. AC motors' superior power density and acceleration characteristics compared to that of DC motors tends to favour PM synchronous, BLDC, induction, and SRM drive applications.

A servo system differs from some stepper motor applications in that the position feedback is continuous while the motor is running; a stepper system relies on the motor not to "miss steps" for short term accuracy, although a stepper system may include a "home" switch or other element to provide long-term stability of control.

For instance, when a typical dot matrix computer printer starts up, its controller makes the print head stepper motor drive to its left-hand limit, where a position sensor defines home position and stops stepping. As long as power is on, a bidirectional counter in the printer's microprocessor keeps track of print-head position.

Stepper Motor

Stepper motors are a type of motor frequently used when precise rotations are required. In a stepper motor an internal rotor containing PMs or a magnetically soft rotor with salient poles is controlled by a set of external magnets that are switched electronically.

A stepper motor may also be thought of as a cross between a DC electric motor and a rotary solenoid. As each coil is energized in turn, the rotor aligns itself with the magnetic field produced by the energized field winding. Unlike a synchronous motor, in its application, the stepper motor may not rotate continuously; instead, it "steps"—starts and then quickly stops again—from one position to the next as field windings are energized and de-energized in sequence.

Depending on the sequence, the rotor may turn forwards or backwards, and it may change direction, stop, speed up or slow down arbitrarily at any time.

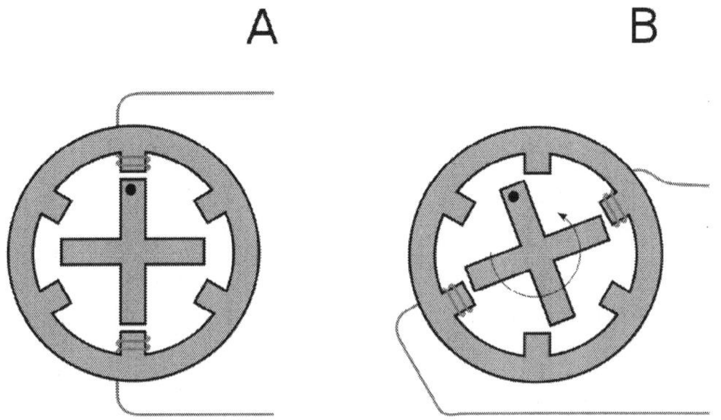

Figure: *A stepper motor with a soft iron rotor, with active windings shown. In 'A' the active windings tend to hold the rotor in position. In 'B' a different set of windings are carrying a current, which generates torque and rotation.*

Simple stepper motor drivers entirely energize or entirely de-energize the field windings, leading the rotor to "cog" to a limited number of positions; more sophisticated drivers can proportionally control the power to the field windings, allowing the rotors to position between the cog points and thereby rotate extremely smoothly.

This mode of operation is often called microstepping. Computer controlled stepper motors are one of the most versatile forms of positioning systems, particularly when part of a digital servo-controlled system.

Stepper motors can be rotated to a specific angle in discrete steps with ease, and hence stepper motors are used for read/write head positioning in computer floppy diskette drives. They were used for the same purpose in pre-gigabyte era computer disk drives, where the precision and speed they offered was adequate for the correct positioning of the read/write head of a hard disk drive.

As drive density increased, the precision and speed limitations of stepper motors made them obsolete for hard drives—the precision limitation made them unusable, and the speed limitation made them uncompetitive—thus newer hard disk drives use voice coil-based head actuator systems. (The term "voice coil" in this connection is historic; it refers to the structure in a typical (cone type) loudspeaker. This structure was used for a while to position the heads.

Modern drives have a pivoted coil mount; the coil swings back and forth, something like a blade of a rotating fan. Nevertheless, like a voice coil, modern actuator coil conductors (the magnet wire) move perpendicular to the magnetic lines of force.)

Stepper motors were and still are often used in computer printers, optical scanners, and digital photocopiers to move the optical scanning element, the print head carriage (of dot matrix and inkjet printers), and the platen or feed rollers.

Likewise, many computer plotters (which since the early 1990s have been replaced with large-format inkjet and laser printers) used rotary stepper motors for pen and platen movement; the typical alternatives here were either linear stepper motors or servomotors with closed-loop analog control systems.

So-called quartz analog wristwatches contain the smallest commonplace stepping motors; they have one coil, draw very little power, and have a PM rotor.

The same kind of motor drives battery-powered quartz clocks. Some of these watches, such as chronographs, contain more than one stepping motor.

Closely related in design to three-phase AC synchronous motors, stepper motors and SRMs are classified as variable reluctance motor type. Stepper motors were and still are often used in computer printers, optical scanners, and computer numerical control (CNC) machines such as routers, plasma cutters and CNC lathes.

Linear Motor

A linear motor is essentially any electric motor that has been "unrolled" so that, instead of producing a torque (rotation), it produces a straight-line force along its length.

Linear motors are most commonly induction motors or stepper motors. Linear motors are commonly found in many roller-coasters where the rapid motion of the motorless railcar is controlled by the rail. They are also used in maglev trains, where the train "flies" over the ground.

On a smaller scale, the 1978 era HP 7225A pen plotter used two linear stepper motors to move the pen along the X and Y axes.

Comparison by Major Categories

Comparison of Motor types

COMPARISON OF MOTOR TYPES				
Type	***Advantages***	***Disadvantages***	***Typical application***	***Typical drive, output***
		Self-commutated motors		
Brushed DC	Simple speed control Low initial cost	Maintenance (brushes) Medium lifespan Costly commutator and brushes	Steel mills Paper making machines Treadmill exercisers Automotive accessories	Rectifier, linear transistor(s) or DC chopper controller.
Brushless DC motor (BLDC) or (BLDM)	Long lifespan Low maintenance High efficiency	Higher initial cost Requires EC controller with closed-loop control	Rigid ("hard") disk drives CD/DVD players Electric vehicles RC Vehicles UAVs	Synchronous; single-phase or three-phase with PM rotor and trapezoidal stator winding; VFD typically VS PWM inverter type.
Switched reluctance motor (SRM)	Long lifespan Low maintenance High efficiency No permanent magnets Low cost Simple construction	Mechanical resonance possible High iron losses Not possible: * Open or vector control * Parallel operation Requires EC controller	Appliances Electric Vehicles Textile mills Aircraft applications	PWM and various other drive types, which tend to be used in very specialized / OEM applications.
Universal motor	High starting torque, compact, high speed.	Maintenance (brushes) Shorter lifespan Usually acoustically noisy Only small ratings are economical	Handheld power tools, blenders, vacuum cleaners, insulation blowers	Variable single phase AC, half-wave or full-wave phase-angle control with triac(s); closed-loop control optional.
		AC asynchronous motors		
AC polyphase squirrel-cage or wound-rotor induction motor (SCIM) or (WRIM)	Self-starting Low cost Robust Reliable Ratings to 1+ MW Standardized types.	High starting current Lower efficiency due to need for magnetization.	Fixed-speed, traditionally, SCIM the world's workhorse especially in low performance applications of all types Variable-speed, traditionally, low-performance variable-torque pumps, fans, blowers and compressors. Variable-speed, increasingly, other high-performance constant-torque and constant-power or dynamic loads.	Fixed-speed, low performance applications of all types. Variable-speed, traditionally, WRIM drives or fixed-speed V/Hz-controlled VSDs. Variable-speed, increasingly, vector-controlled VSDs displacing DC, WRIM and single-phase AC induction motor drives.

AC SCIM SPLIT-PHASE CAPACITOR-START	HIGH POWER HIGH STARTING TORQUE	SPEED SLIGHTLY BELOW SYNCHRONOUS STARTING SWITCH OR RELAY REQUIRED	APPLIANCES STATIONARY POWER TOOLS	FIXED OR VARIABLE SINGLE-PHASE AC, VARIABLE SPEED BEING DERIVED, TYPICALLY, BY FULL-WAVE PHASE-ANGLE CONTROL WITH TRIAC(S); CLOSED-LOOP CONTROL OPTIONAL.
AC SCIM split-phase capacitor-run	Moderate power High starting torque No starting switch Comparatively long life	Speed slightly below synchronous Slightly more costly	Industrial blowers Industrial machinery	
AC SCIM split-phase, auxiliary start winding	Moderate power Low starting torque	Speed slightly below synchronous Starting switch or relay required	Appliances Stationary power tools	
AC induction shaded-pole motor	Low cost Long life	Speed slightly below synchronous Low starting torque Small ratings low efficiency	Fans, appliances, record players	

AC SYNCHRONOUS MOTORS				
Wound-rotor synchronous motor (WRSM)	Synchronous speed Inherently more efficient induction motor, low power factor	More costly	Industrial motors	Fixed or variable speed, three-phase; VFD typically six-step CS load-commutated inverter type or VS PWM inverter type.
Hysteresis motor	Accurate speed control Low noise No vibration High starting torque	Very low efficiency	Clocks, timers, sound producing or recording equipment, hard drive, capstan drive	Single-phase AC, two-phase capacitor-start, capacitor run motor
Synchronous reluctance motor (SyRM)	Equivalent to SCIM except more robust, more efficient, runs cooler, smaller footprint Competes with PM synchronous motor without demagnetization issues	Requires a controller Not widely available High cost	Appliances Electric vehicles Textile mills Aircraft applications	VFD can be standard DTC type or VS inverter PWM type.

SPECIALITY MOTORS				
Pancake or axial rotor motors	Compact design Simple speed control	Medium cost Medium lifespan	Office Equip Fans/Pumps, fast industrial and military servos	Drives can typically be brushed or brushless DC type.
Stepper motor	Precision positioning High holding torque	Some can be costly Require a controller	Positioning in printers and floppy disc drives; industrial machine tools	Not a VFD. Stepper position is determined by pulse counting.

Electromagnetism

Force and Torque

The fundamental purpose of the vast majority of the world's electric motors is to electromagnetically induce relative movement in an air gap between a stator and rotor to produce useful torque or linear force.

According Lorentz force law the force of a winding conductor can be given simply by:

$$\mathrm{F} = I\ell \times \mathrm{B}$$

or more generally, to handle conductors with any geometry:

$$\mathrm{F} = \mathrm{J} \times \mathrm{B}$$

The most general approaches to calculating the forces in motors use tensors.

Power

Where rpm is shaft speed and T is torque, a motor's mechanical power output P_{em} is given by,

in British units with T expressed in foot-pounds,

$$P_{em} = \frac{rpm \times T}{5252} \text{ (horsepower), and,}$$

in SI units with shaft speed expressed in radians per second, and T expressed in newton-metres,

$$P_{em} = speed \times T \text{ (watts).}$$

For a linear motor, with force F expressed in newtons and velocity v expressed in metres per second,

$$P_{em} = F \times v \text{ (watts).}$$

In an asynchronous or induction motor, the relationship between motor speed and air gap power is, neglecting skin effect, given by the following:

$$P_{airgap} = \frac{R_r}{s} * I_r^2 \text{, where}$$

R_r - rotor resistance

I_r^2 - square of current induced in the rotor

s - motor slip; ie, difference between synchronous speed and slip speed, which provides the relative movement needed for current induction in the rotor.

Back Emf

Since the armature windings of a direct-current motor are moving through a magnetic field, they have a voltage induced in them. This voltage tends to oppose the motor supply voltage and so is called "back electromotive force (emf)". The voltage is proportional to the running speed of the motor. The back emf of the motor, plus the voltage drop across the winding internal resistance and brushes, must equal the voltage at the brushes. This provides the fundamental mechanism of speed regulation in a DC motor. If the mechanical load increases, the motor slows down; a lower back emf results, and more current is drawn from the supply. This increased current provides the additional torque to balance the new load.

In AC machines, it is sometimes useful to consider a back emf source within the machine; this is of particular concern for close speed regulation of induction motors on VFDs, for example.

Losses

Motor losses are mainly due to resistive losses in windings, core losses and mechanical losses in bearings, and aerodynamic losses, particularly where cooling fans are present, also occur.

Losses also occur in commutation, mechanical commutators spark, and electronic commutators and also dissipate heat.

Efficiency

To calculate a motor's efficiency, the mechanical output power is divided by the electrical input power:

$$\eta = \frac{P_m}{P_e},$$

where η is energy conversion efficiency, P_e is electrical input power, and P_m is mechanical output power:

$$P_e = IV$$

$$P_m = T\omega$$

where V is input voltage, I is input current, T is output torque, and ω is output angular velocity. It is possible to derive analytically the point of maximum efficiency. It is typically at less than 1/2 the stall torque.

Various regulatory authorities in many countries have introduced and implemented legislation to encourage the manufacture and use of higher efficiency electric motors.

Goodness Factor

Professor Eric Laithwaite proposed a metric to determine the 'goodness' of an electric motor: $G = \frac{\omega}{resistance \times reluctance} = \frac{\omega \mu \sigma A_m A_e}{l_m l_e}$

Where:

G is the goodness factor (factors above 1 are likely to be efficient)

A_m, A_e are the cross sections of the magnetic and electric circuit

l_m, l_e are the lengths of the magnetic and electric circuits

μ is the permeability of the core

ω is the angular frequency the motor is driven at

From this, he showed that the most efficient motors are likely to have relatively large magnetic poles. However, the equation only directly relates to non PM motors.

Performance Parameters

Torque Capability of Motor Types

All the electromagnetic motors, and that includes the types mentioned here derive the torque from the vector product of the interacting fields. For calculating the torque it is necessary to know the fields in the air gap. Once these have been established by mathematical analysis using FEA or other tools the torque may be calculated as the integral of all the vectors of force multiplied by the radius of each vector. The current flowing in the windings is producing the fields and for a motor using a magnetic material the field is not linearilly proportional to the current. This makes the calculation difficult but a computer can do the many calculations needed. Once this is done a figure relating the current to the torque can be used as a useful parameter for motor selection.

The maximum torque for a motor will depend on the maximum current although this will usually be .only usable until thermal considerations take precedence. When optimally designed within a given core saturation constraint and for a given active current (i.e., torque current), voltage, pole-pair number, excitation frequency (i.e., synchronous speed), and air-gap flux density, all categories of electric motors or generators will exhibit virtually the same maximum continuous shaft torque (i.e., operating torque) within a given air-gap area with winding slots and back-iron depth, which determines the physical size of electromagnetic core. Some applications require bursts of torque beyond the maximum operating torque, such as short bursts of torque to accelerate an electric vehicle from standstill. Always limited by magnetic core saturation or safe operating temperature rise and voltage, the capacity for torque bursts beyond the maximum operating torque differs significantly between categories of electric motors or generators.

Capacity for bursts of torque should not be confused with field weakening capability. Field weakening allows an electric machine to operate beyond the designed frequency of excitation. Field weakening is done when the maximum speed cannot be reached by increasing the applied voltage. This applies to only motors with current controlled fields and therefore cannot be achieved with PM motors.

Electric machines without a transformer circuit topology, such as that of WRSMs or PMSMs, cannot realise bursts of torque higher than the maximum designed torque without saturating the magnetic core and rendering any increase in current as useless. Furthermore, the PM assembly of PMSMs can be irreparably damaged, if bursts of torque exceeding the maximum operating torque rating are attempted.

Electric machines with a transformer circuit topology, such as induction machines, induction doubly fed electric machines, and induction or synchronous wound-rotor doubly fed (WRDF) machines, exhibit very high bursts of torque because the emf-induced active current on either side of the transformer oppose each other and thus contribute nothing to the transformer coupled magnetic core flux density, which would otherwise lead to core saturation.

Electric machines that rely on induction or asynchronous principles short-circuit one port of the transformer circuit and as a result, the reactive impedance of the transformer circuit becomes dominant as slip increases, which limits the magnitude of active (i.e., real) current. Still, bursts of torque that are two to three times higher than the maximum design torque are realisable.

The brushless wound-rotor synchronous doubly fed (BWRSDF) machine is the only electric machine with a truly dual ported transformer circuit topology (i.e., both ports independently excited with no short-circuited port). The dual ported transformer circuit topology is known to be unstable and requires a multiphase slip-ring-brush assembly to propagate limited power to the rotor winding set. If a precision means were available to instantaneously control torque angle and slip for synchronous operation during motoring or generating while simultaneously providing brushless power to the rotor winding set, the active current of the BWRSDF machine would be independent of the reactive impedance of the transformer circuit and bursts of torque significantly higher than the maximum operating torque and far beyond the practical capability of any other type of electric machine would be realisable. Torque bursts greater than eight times operating torque have been calculated.

Continuous Torque Density

The continuous torque density of conventional electric machines is determined by the size of the air-gap area and the back-iron depth, which are determined by the power rating of the armature winding set, the speed of the machine, and the achievable air-gap flux density before core saturation. Despite the high coercivity of neodymium or samarium-cobalt PMs, continuous torque density is virtually the same amongst electric machines with optimally designed armature winding sets. Continuous torque density relates to method of cooling and permissible period of operation before destruction by overheating of windings or PM damage.

Continuous Power Density

The continuous power density is determined by the product of the continuous torque density and the constant torque speed range of the electric machine.

Standards

The following are major design and manufacturing standards covering electric motors:

- International Electrotechnical Commission: IEC 60034 Rotating Electrical Machines
- National Electrical Manufacturers Association: MG-1 Motors and Generators
- Underwriters Laboratories: UL 1004 - Standard for Electric Motors

Non-Magnetic Motors

An electrostatic motor is based on the attraction and repulsion of electric charge. Usually, electrostatic motors are the dual of conventional coil-based motors. They typically require a high voltage power supply, although very small motors employ lower voltages. Conventional electric motors instead employ magnetic attraction and repulsion, and require high current at low voltages. In the 1750s, the first electrostatic motors were developed by Benjamin Franklin and

Andrew Gordon. Today the electrostatic motor finds frequent use in micro-electro-mechanical systems (MEMS) where their drive voltages are below 100 volts, and where moving, charged plates are far easier to fabricate than coils and iron cores. Also, the molecular machinery which runs living cells is often based on linear and rotary electrostatic motors.

A piezoelectric motor or piezo motor is a type of electric motor based upon the change in shape of a piezoelectric material when an electric field is applied. Piezoelectric motors make use of the converse piezoelectric effect whereby the material produces acoustic or ultrasonic vibrations in order to produce a linear or rotary motion. In one mechanism, the elongation in a single plane is used to make a series stretches and position holds, similar to the way a caterpillar moves.

An electrically powered spacecraft propulsion system uses electric motor technology to propel spacecraft in outer space, most systems being based on electrically powering propellant to high speed, with some systems being based on electrodynamic tethers principles of propulsion to the magnetosphere.

Motor Capacitor

A motor capacitor, such as a start capacitor or run capacitor (including a dual run capacitor) is an electrical capacitor that alters the current to one or more windings of a single phase AC induction motor to create a rotating magnetic field. There are two common.types of motor capacitors, *run capacitors* and *start capacitors*. The units of capacitance are labelled in microfarads (μF or uF or "mfd" or "MFD", which still refers to "micro", not "milli").

Motor capacitors are used with air conditioners, hot tub/jacuzzi spa pumps, or forced air heat furnaces for example. A "dual run capacitor" is used in some air conditioner compressor units, to boost both the fan and compressor motors.

Run Capacitors

Some single-phase AC electric motors require a "run capacitor" to energize the second-phase winding (auxiliary coil) to create a rotating magnetic field while the motor is running.

Run capacitors are designed for continuous duty while the motor is powered, which is why electrolytic capacitors are avoided, and low-loss polymer capacitors are used instead. Run capacitors are mostly polypropylene film capacitors and are designed for continuous duty; they are energized the entire time the motor is running. Run capacitors are rated in a range of 1.5 to 100 microfarads (µF or mfd), with voltage classifications of 370 V or 440 V.

If a wrong capacitance value is installed, it will cause an uneven magnetic field which causes the rotor to hesistate at those spots that are uneven, which can be observed as uneven motor rotation speed, especially under load. This hesitation can cause the motor to become noisy, increase energy consumption, cause performance to drop, and cause the motor to overheat.

Start Capacitors

Start capacitors briefly increase motor starting torque and allow a motor to be cycled on and off rapidly. A start capacitor stays in the circuit long enough to rapidly bring the motor up to a predetermined speed, which is usually about 75% of the full speed, and is then taken out of the circuit, such as by a centrifugal switch that releases at that speed, afterward the motor works more efficiently with a run capacitor.

Start capacitors have ratings above 70 microfarads (µF), with four major voltage classifications: 125 V, 165 V, 250 V, and 330 V. Examples of motor capacitors are: a 35 µF/mfd, at 370 V, run capacitor, or an 88–108 µF/mfd at 250 V start capacitor.

Start capacitors above 20 microfarad (µF) are always non-polarized aluminium electrolytic capacitors with non solid electrolyte and therefore they are only applicable for the short motor starting time.

The motor won't work properly if the centrifugal switch is broken. If the switch is always "open", the start capacitor is not part of the circuit thus preventing startup of the motor. If the switch is always "closed", the start capacitor is always enabled, thus likely destroying the capacitor. If a motor doesn't start, the capacitor is far more likely the problem instead of the switch.

Dual Run Capacitors

A dual "run capacitor" supports two electric motors, such as in large air conditioner or heat pump units, with both a fan motor and a compressor motor. It saves space by combining two physical capacitors into one case. The dual capacitor has 3 terminals labelled "C", "FAN", and "HERM", which stand for the Common, Fan, and HERMetically sealed compressor.

Dual capacitors come in a variety of sizes, depending on the capacitance (µF), such as 40 plus 5 µF, and also the voltage. A 440 volt capacitor can be used in place of a 370 volt, but not a 370 in place of a 440 volt. The capacitance must stay the same within 5% of its original value. Round cylinder-shaped dual run capacitors are commonly used for air conditioning, to help in the starting of the compressor and the condenser fan motor. An oval dual run capacitor could be used instead of a round capacitor, but the mounting strap should be changed to better fit the oval shape.

Failure Modes

A faulty run capacitor often becomes swollen, with the sides or ends bowed or bulged out further than usual: it can be clear to see that the capacitor has failed because it is swollen or even blown apart with capacitor oil leaking. Some capacitors are built with a "Pressure Sensitive Interrupter" design that causes them to fail before internal pressures can cause serious injury. One design causes the top of the capacitor to expand and break internal wiring.

"Weak Capacitor" - Over many years of use the capacitance of the capacitor (measured in microfarads) reduces. As a result the motor may fail to start or run at full power.

If a motor is running during a lightning strike on the power grid, the run capacitor might be damaged or weaken by a voltage spike, thus requiring replacement.

IEC/EN 60252-1 2001 specifies levels of protection for motor run capacitors

- P0 - no protection;
- P1 - fail open circuit or short circuit
- P2 - fail open circuit only

Safety Issues

The motor capacitor, which is a component of a jacuzzi circulating pump, can overheat if defective. This poses a fire hazard, and the U.S. Consumer Product Safety Commission (CPSC) has received more than 100 reports of incidents of overheating of the motor capacitor, with some fires started.

ICT in Agriculture

The application of information and communications technology (ICT) in agriculture is increasingly important.

E-Agriculture is an emerging field focusing on the enhancement of agricultural and rural development through improved information and communication processes. More specifically, e-Agriculture involves the conceptualization, design, development, evaluation and application of innovative ways to use information and communication technologies

(IT) in the rural domain, with a primary focus on agriculture. E-Agriculture is a relatively new term and we fully expect its scope to change and evolve as our understanding of the area grows.

E-Agriculture is one of the action lines identified in the declaration and plan of action of the World Summit on the Information Society (WSIS). The "Tunis Agenda for the Information Society," published on 18 November 2005 and emphasizes the leading facilitating roles that UN agencies need to play in the implementation of the Geneva Plan of Action. The Food and Agriculture Organisation of the United Nations (FAO) has been assigned the responsibility of organising activities related to the action line under C.7 ICT Applications on E-Agriculture.

The main phases of the agriculture industry include crop cultivation, water management, fertilizer application, fertigation, pest management, harvesting, post-harvest handling, transport of food products, packaging, food preservation, food processing/value addition, quality management, food safety, food storage, and food marketing.

All stakeholders of agriculture industry need information and knowledge about these phases to manage them efficiently. Any system applied for getting information and knowledge for making decisions in any industry should deliver accurate, complete, concise information in time or on time. The information provided by the system must be in user-friendly form, easy to access, cost-effective and well protected from unauthorized accesses.

- Record text, drawings, photographs, audio, video, process descriptions, and other information in digital formats,
- Produce exact duplicates of such information at significantly lower cost,
- Transfer information and knowledge rapidly over large distances through communications networks.
- Develop standardized algorithms to large quantities of information relatively rapidly.
- Achieve greater interactivity in communicating, evaluating, producing and sharing useful information and knowledge.

The main focus of this article is to elaborate how the achievements of IT can be applied in Agriculture sector and its development. The main applications of ICT in Agriculture sector are listed below.

Office Automation

The office automation is application of computers, computer networks, telephone networks, and other office automation tool such as photocopy machines, scanners, printers, cleaning equipment, and electronic security systems to increase the productivity of organisations.

There are many government, private and non-government organisations involved in agriculture sector and rural development. They all have to work together to give better service to farming community. Therefore, application of office automation is one of the solutions to enhance the efficiency and inter-connectivity of the employees work in all above mentioned organisations.

Many computer applications such as MS Office, Internet Explorer, OpenOffice.org and other tailor-made office automation software packages are providing unlimited potential to organisations and individuals to fulfill their day to day data processing requirements to give an efficient service to their customers.

Wireless Technologies

Wireless technologies have numerous applications in agriculture. One major usage is the simplification of closed-circuit television camera systems; the use of wireless communications eliminates the need for the installation of coaxial cables.

Global Positioning System

In agriculture, the use of the Global Positioning System provides benefits in geo-fencing, map-making and surveying. GPS receivers dropped in price over the years, making it more popular for civilian use. With the use of GPS, civilians can produce simple yet highly accurate digitized map without the help of a professional cartographer. In Kenya, for example, the solution to prevent an elephant bull from wandering into farms and destroying precious crops was to tag the elephant with a device that sends a text message when it crosses a geo-fence. Using the technology of SMS and GPS, the elephant can roam freely and the authorities are alerted whenever it is near the farm.

Geographic Information Systems

Geographic information system s, or GIS, are extensively used in agriculture, especially in precision farming. Land is mapped digitally, and pertinent geodetic data such as topography and contours are combined with other statistical data for easier analysis of the soil. GIS is used in decision making such as what to plant and where to plant using historical data and sampling.

Computer-Controlled Devices (Automated Systems)

Automatic milking systems are computer controlled stand alone systems that milk the dairy cattle without human labour. The complete automation of the milking process is controlled by an agricultural robot, a complex herd management software, and specialised computers. Automatic milking eliminates the farmer from the actual milking process, allowing for more time for supervision of the farm and the herd. Farmers can also improve herd management by using the data gathered by the computer. By analyzing the effect of various animal feeds on milk yield, farmers may adjust accordingly to obtain optimal milk yields. Since the data is available down to individual level, each cow may be tracked and examined, and the farmer may be alerted when there are unusual changes that could mean sickness or injuries.

RFID

The Veterinary Department of Malaysia's Ministry of Agriculture introduced a livestock-tracking program in 2009 to track the estimated 80,000 cattle all across the country. Each cattle is tagged with the use of RFID technology for easier identification, providing access to relevant data such as: bearer's location, name of breeder, origin of livestock, sex, and dates of movement. This program is the first of its kind in Asia, and is expected to increase the competitiveness of Malaysian livestock industry in international markets by satisfying the regulatory requirements of importing countries like United States, Europe and Middle East. Tracking by RFID will also help producers meet the dietary standards by the halal market. The program will also provide improvements in controlling disease outbreaks in livestock.

Internal Combustion Engine

An internal combustion engine (ICE) is an engine where the combustion of a fuel occurs with an oxidizer (usually air) in a combustion chamber that is an integral part of the working fluid flow circuit. In an internal combustion engine the expansion of the high-temperature and high-pressure gases produced by combustion apply direct force to some component of the engine. The force is applied typically to pistons, turbine blades, or a nozzle. This force moves the component over a distance, transforming chemical energy into useful mechanical energy. The first commercially successful internal combustion engine was created by Étienne Lenoir around 1859.

The term *internal combustion engine* usually refers to an engine in which combustion is intermittent, such as the more familiar four-stroke and two-stroke piston engines, along with variants, such as the six-stroke piston engine and the Wankel rotary engine. A second class of internal combustion engines use continuous combustion: gas turbines, jet engines and most rocket engines, each of which are internal combustion engines on the same principle as previously described. Firearms are also a form of internal combustion engine.

Internal combustion engines are quite different from external combustion engines, such as steam or Stirling engines, in which the energy is delivered to a working fluid not consisting of, mixed with, or contaminated by combustion products. Working fluids can be air, hot water, pressurized water or even liquid sodium, heated in a boiler. ICEs are usually powered by energy-dense fuels such as gasoline or diesel, liquids derived from fossil fuels. While there are many stationary applications, most ICEs are used in mobile applications and are the dominant power supply for cars, aircraft, and boats.

Typically an ICE is fed with fossil fuels like natural gas or petroleum products such as gasoline, diesel fuel or fuel oil. There's a growing usage of renewable fuels like biodiesel for compression ignition engines and bioethanol for spark ignition engines. Hydrogen is sometimes used, and can be made from either fossil fuels or renewable energy.

Applications

Reciprocating piston engines are by far the most common power source for land vehicles including automobiles, motorcycles, locomotives and ships. Wankel engines are found on some automobiles and motorcycles.

Where very high power-to-weight ratios are required, internal combustion engines appear in the form of combustion turbines. Powered aircraft typically use an ICE which may be a reciprocating engine. Airplanes can also use jet engines and helicopters can also employ turboshafts; both of these are types of turbines. In addition to providing propulsion, airliners employ a separate ICE as an auxiliary power unit.

ICEs also have a role in large scale electric power generation where they're found in the form of combustion turbines in combined cycle power plants with typical electrical output in the range of 100 MW to 1 GW. The high temperature exhaust is used to boil and superheat water to run a steam turbine. Thus, more energy is extracted from the fuel that could be extracted by the combustion turbine alone. In combined cycle power plants efficiencies in the range of 50 % to 60 % are typical. In a smaller scale Diesel generators are used for backup power and for providing electrical power to areas not connected to an electric grid.

Two-stroke engines are widely used in snowmobiles, lawnmowers, string trimmers, chain saws, jet skis, mopeds, outboard motors, and many motorcycles.

Two stroke gasoline small engines are a common power source for chainsaws, leafblowers and lawnmowers.

Engine Configurations

Internal combustion engines can be classified by their configuration.

Common layouts of engines are:

Reciprocating:

- Two-stroke engine
- Four-stroke engine (Otto cycle)
- Six-stroke engine
- Diesel engine
- Atkinson cycle
- Miller cycle

Rotary:

- Wankel engine

Continuous combustion:

- Gas turbine
- Jet engine (including turbojet, turbofan, ramjet, rocket, etc.)

Reciprocating Engines

Structure

The base of a reciprocating internal combustion engine is the engine block which is typically made of cast iron or aluminium. The engine block contains the cylinders. In engines with more than 1 cylinder they're usually arranged either in 1 row (straight engine) or 2 rows (boxer engine or V engine); 3 rows are occasionally used (W engine) in contemporary engines, and other engine configurations are possible and have been used. Single cylinder engines are common for motorcycles and in small engines of machinery. Water cooled engines contain passages in the engine block where cooling fluid circulate (the water jacket). Some small engines are air cooled, and instead of having a water jacket the cylinder block has fins protruding away from it to cool by directly transferring heat to the air. The cylinder walls are usually finished by honing to obtain a cross hatch which is better able to retain the oil. A too rough surface would quickly harm the engine by excessive wear on the piston.

The pistons are short cylindrical parts which seal one end of the cylinder from the high pressure of the compressed air and combustion products and slide continuously within it while the engine is in operation. The top wall of the piston is termed its *crown* and is typically flat or concave. Some two stroke engines use pistons with a deflector head. Pistons are open at the bottom and hollow except for an integral reinforcement structure (the piston web). When an engine is working the gas pressure in the combustion chamber exerts a force on the piston crown which is transferred through its web to a gudgeon pin. Each piston has rings fitted around its circumference that mostly prevent the gases from leaking into the crankcase or the oil into the combustion chamber. A ventilation system drives the small amount of gas that escape past the pistons during normal operation (the blow-by gases) out of the crankcase so that it does not accumulate contaminating the oil and creating corrosion. In two stroke gasoline engines the crankcase is part of the air–fuel path and due to the continuous flow of it they do not need a separate crankcase ventilation system.

The cylinder head is attached to the engine block by numerous bolts or studs. It has several functions. The cylinder head seals the cylinders on the side opposite to the pistons; it contains short ducts (the *ports*) for intake and exhaust and the associated intake valves that open to let the cylinder be filled with fresh air and exhaust valves that open to allow the combustion gases to escape. However, 2-stroke crankcase scavenged engines connect the gas ports directly to the cylinder wall without poppet valves; the piston controls their opening and occlusion instead. The cylinder head also holds the spark plug in the case of spark ignition engines and the injector for engines that use direct injection. All CI engines use fuel injection, usually direct injection but some engines instead use indirect injection. SI engines can use a carburetor or fuel injection as port injection or direct injection. Most SI engines have a single spark plug per cylinder but some have 2. A head gasket prevents the gas from leaking between the cylinder head and the engine block. The opening and closing of the valves is controlled by one or several camshafts and springs—or in some engines—a desmodromic mechanism that uses no springs. The camshaft may press directly the stem of the valve or may act upon a rocker arm, again, either directly or through a pushrod.

The crankcase is sealed at the bottom with a sump that collects the falling oil during normal operation to be cycled again. The cavity created between the cylinder block and the sump houses a crankshaft that converts the reciprocating motion of the pistons to rotational motion. The crankshaft is held in place relative to the engine block by main bearings, which allow it to rotate. Bulkheads in the crankcase form a half of every main bearing; the other half is a detachable cap. In some cases a single *main bearing deck* is used rather than several smaller caps. A connecting rod is connected to offset sections of the crankshaft (the crankpins) in one end and to the piston in the other end through the gudgeon pin and thus transfers the force and translates the reciprocating motion of the pistons to the circular motion of the crankshaft. The end of the connecting rod attached to the gudgeon pin is called its small end, and the other end, where it is connected to the crankshaft, the big end. The big end has a detachable half to allow assembly around the crankshaft. It is kept together to the connecting rod by removable bolts.

The cylinder head has attached an intake manifold an exhaust manifold to the corresponding ports. The intake manifold connects to the air filter directly, or to a carburetor when one is present, which is then connected to the air filter. It distributes the air incoming from these devices to the individual cylinders. The exhaust manifold is the first component in the exhaust system. It collects the exhaust gases from the cylinders and drives it to the following component in the path. The exhaust system of an ICE may also include a catalytic converter and muffler. The final section in the path of the exhaust gases is the tailpipe.

4-Stroke Engines

The *top dead centre* (TDC) of a piston is the position where it is nearest to the valves; *bottom dead centre* (BDC) is the opposite position where it is furtherest from them. A *stroke* is the movement of a piston from TDC to BDC or vice-versa together with the associated process. While an engine is in operation the crankshaft rotates continuously at a nearly constant speed. In a 4-stroke ICE each piston experiments 2 strokes per crankshaft revolution in the following order. Starting the description at TDC, these are:

1. Intake, induction or suction: The intake valves are open as a result of the cam lobe pressing down on the valve stem. The piston moves downward increasing the volume of the combustion chamber and allowing air to enter in the case of a CI engine or an air fuel mix in the case of SI engines that do not use direct injection. The air or air-fuel mixture is called the *charge* in any case.

2. Compression: In this stroke, both valves are closed and the piston moves upward reducing the combustion chamber volume which reaches its minimum when the piston is at TDC. The piston performs work on the charge as it is being compressed; as a result its pressure, temperature and density increase; an approximation to this behaviour is provided by the ideal gas law. Just before the piston reaches TDC, ignition begins. In the

case of a SI engine, the spark plug receives a high voltage pulse that generates the spark which gives it its name and ignites the charge. In the case of a CI engine the fuel injector quickly injects fuel into the combustion chamber as a spray; the fuel ignites due to the high temperature.

3. Power: The pressure of the combustion gases pushes the piston downward exerting more work than it was made to compress the charge. Complementary to the compression stroke, the combustion gases expand and as a result their temperature, pressure and density decreases. When the piston is near to BDC the exhaust valve opens. The combustion gases expand irreversibly due to the leftover pressure—in excess of back pressure, the gauge pressure on the exhaust port—, this is called the *blowdown*.
4. Exhaust: The exhaust valve remains open while the piston moves upward expelling the combustion gases. For naturally aspirated engines a small part of the combustion gases may remain in the cylinder during normal operation because the piston does not close the combustion chamber completely; these gases dissolve in the next charge. At the end of this stroke, the exhaust valve closes, the intake valve opens, and the sequence repeats in the next cycle. The intake valve may open before the exhaust valve closes to allow better scavenging.

2-Stroke Engines

The defining characteristic of this kind of engine is that each piston completes a cycle every crankshaft revolution. The 4 process of intake, compression, power and exhaust take place in only 2 strokes so that it is not possible to dedicate a stroke exclusively for each of them. Starting at TDC the cycle consist of:

1. Power: While the piston is descending the combustion gases perform work on it—as in a 4-stroke engine—. The same thermodynamical considerations about the expansion apply.
2. Scavenging: Around 75° of crankshaft rotation before BDC the exhaust valve or port opens, and blowdown occurs. Shortly thereafter the intake valve or transfer port opens. The incoming charge displaces the remaining combustion gases to the exhaust system and a part of the charge may enter the exhaust system as well. The piston reaches BDC and reverses direction. After the piston has travelled a short distance upwards into the cylinder the exhaust valve or port closes; shortly the intake valve or transfer port closes as well.

3. Compression: With both intake and exhaust closed the piston continues moving upwards compressing the charge and performing a work on it. As in the case of a 4-stroke engine, ignition starts just before the piston reaches TDC and the same consideration on the thermodynamics of the compression on the charge.

While a 4-stroke engine uses the piston as a positive displacement pump to accomplish scavenging taking 2 of the 4 strokes, a 2-stroke engine uses the last part of the power stroke and the first part of the compression stroke for combined intake and exhaust. The work required to displace the charge and exhaust gases comes from either the crankcase or a separate blower. For scavenging, expulsion of burned gas and entry of fresh mix, two main approaches are described: 'Loop scavenging', and 'Uniflow scavenging', SAE news published in the 2010s that 'Loop Scavenging' is better under any circumstance than 'Uniflow Scavenging'.

Crankcase Scavenged

Some SI engines are crankcase scavenged and do not use poppet valves. Instead the crankcase and the part of the cylinder below the piston is used as a pump. The intake port is connected to the crankcase through a reed valve or a rotary disk valve driven by the engine. For each cylinder a transfer port connects in one end to the crankcase and in the other end to the cylinder wall. The exhaust port is connected directly to the cylinder wall. The transfer and exhaust port are opened and closed by the piston. The reed valve opens when the crankcase is slightly below intake pressure to let it be filled with a new charge, this happens when the piston is moving upwards. When the piston is moving downwards the pressure in the crankcase increases and the reed valve closes promptly, then the charge in the crankcase is compressed. When the piston uncovers the transfer port the higher pressure of the charge in the crankcase makes it enter the cylinder, blowing the exhaust gases. Lubrication is accomplished by adding *2-stroke oil* to the fuel in small ratios. *Petroil* refers to the mix of gasoline with the aforesaid oil. This kind of 2-stroke engines has a lower efficiency than comparable 4-strokes engines and release a more polluting exhaust gases for the following conditions:

- They use a *total-loss lubrication system*: all the lubricating oil is eventually burned along with the fuel.
- There are conflicting requirements for scavenging: On one side, enough fresh charge needs to be introduced in each cycle to displace almost all the combustion gases but too much of it means that a part of it gets in the exhaust.

- They must use the transfer port(s) as a carefully designed and placed nozzle so that a gas current is created in a way that it sweeps the whole cylinder before reaching the exhaust port. 4-stroke engines have the benefit of forcibly expelling almost all of the combustion gases, and during exhaust the combustion chamber is reduced to a minimum volume. In crankcase scavenged 2-stroke engines, exhaust and intake are performed mostly simultaneously and with the combustion chamber at the maximum volume.

The main advantage of 2-stroke engines of this type is mechanical simplicity and a higher power-to-weight ratio than their 4-stroke counterparts. Despite having twice as many power strokes per cycle, less than twice the power of a comparable 4-stroke engine is attainable in practice.

Blower Scavenged

Using a separate blower avoids many of the shortcomings of crankcase scavenging, at the expense of increased complexity which means an higher cost and an increase in maintenance requirement. An engine of this type uses ports or valves for intake and valves for exhaust except opposed piston engines, which may also use ports for exhaust. The blower is usually of the Roots-type but other types have been used too. This design is commonplace in CI engines, and has been occasionally used in SI engines.

CI engines that use a blower typically use *uniflow scavenging*. In this design the cylinder wall contains several intake ports just above the position that the piston crown reaches when at BDC. An exhaust valve or several like that of 4-stroke engines is used. The final part of the intake manifold is an air sleeve which feeds the intake ports. The intake ports are placed at an horizontal angle to the cylinder wall (I.e: they are in plane of the piston crown) to give a swirl to the incoming charge to improve combustion. The largest reciprocating IC are low speed CI engines of this type; they are used for marine propulsion or electric power generation and achieve the highest thermal efficiencies among internal combustion engines of any kind. Some Diesel-electric locomotive engines operate on the 2-stroke cycle. The most powerful of them have a brake power of around 4.5 MW or 6,000 HP. The EMD SD90MAC class of locomotives use a 2-stroke engine. The comparable class GE AC6000CW whose prime mover has almost the same brake power uses a 4-stroke engine.

An example of this type of engine is the Wärtsilä-Sulzer RTA96-C turbocharged 2-stroke Diesel, used in large container ships. It is the most efficient and powerful internal combustion engine in the world with a thermal efficiency over 50 %. For comparison, the most efficient small four-stroke engines are around 43 % thermally-efficient (SAE 900648); size is an advantage for efficiency due to the increase in the ratio of volume to surface area.

Ignition

Internal combustion engines require ignition of the mixture, either by spark ignition (SI) or compression ignition (CI). Before the invention of reliable electrical methods, hot tube and flame methods were used. Experimental engines with laser ignition have been built.

Gasoline Ignition Process

Gasoline engine ignition systems generally rely on a combination of alternator or generator and lead–acid battery for electrical power. The battery supplies electrical power for cranking, and supplies electrical power when the engine is off. The battery also supplies electrical power during rare run conditions where the alternator cannot maintain more than 13.8 volts (for a common 12V automotive electrical system). As alternator voltage falls below 13.8 volts, the lead-acid storage battery increasingly picks up electrical load. During virtually all running conditions, including normal idle conditions, the alternator supplies primary electrical power.

Some systems disable alternator field (rotor) power during wide open throttle conditions. Disabling the field reduces alternator pulley mechanical loading to nearly zero, maximizing crankshaft power. In this case the battery supplies all primary electrical power.

Gasoline engines take in a mixture of air and gasoline and compress it to not more than 12.8 bar (1.28 MPa). When mixture is compressed, as the piston approaches the cylinder head and maximum stroke, a spark plug ignites the mixture.

The necessary high voltage, typically 10,000 volts to over 30,000 volts, is supplied by an induction coil or transformer. The induction coil is a fly-back system, using interruption of electrical primary system current through some type of synchronized interrupter. The interrupter can be either contact points or a power transistor. Some ignition systems are capacitive discharge types. CD ignitions use step-up transformers. The step-up transformer uses energy stored in a capacitance to generate electric spark. With either system, a mechanical or electrical control system provides a carefully timed high-voltage to the proper cylinder. This spark, via the spark plug, ignites the air-fuel mixture in the engine's cylinders.

While gasoline internal combustion engines are much easier to start in cold weather than diesel engines, they can still have cold weather starting problems under extreme conditions. For years the solution was to park the car in heated areas. In some parts of the world the oil was actually drained and heated over night and returned to the engine for cold starts. In the early 1950s the gasoline Gasifier unit was developed, where, on cold weather starts, raw gasoline was diverted to the unit where part of the fuel was burned causing the other part to become a hot vapor sent directly to the intake valve manifold. This unit was quite popular until electric engine block heaters became standard on gasoline engines sold in cold climates.

Diesel Ignition Process

Diesel engines and HCCI (Homogeneous charge compression ignition) engines, rely solely on heat and pressure created by the engine in its compression process for ignition. The compression level that occurs is usually twice or more than a gasoline engine. Diesel engines take in air only, and shortly before peak compression, spray a small quantity of diesel fuel into the cylinder via a fuel injector that allows the fuel to instantly ignite. HCCI type engines take in both air and fuel, but continue to rely on an unaided auto-combustion process, due to higher pressures and heat. This is also why diesel and HCCI engines are more susceptible to cold-starting issues, although they run just as well in cold weather once started. Light duty diesel engines with indirect injection in automobiles and light trucks employ glowplugs that pre-heat the combustion chamber just before starting to reduce no-start conditions in cold weather. Most diesels also have a battery and charging system; nevertheless, this system is secondary and is added by manufacturers as a luxury for the ease of starting, turning fuel on and off (which can also be done via a switch or mechanical apparatus), and for running auxiliary electrical components and accessories. Most new engines rely on electrical and electronic engine control units (ECU) that also adjust the combustion process to increase efficiency and reduce emissions.

Lubrication

Surfaces in contact and relative motion to other surfaces require lubrication to reduce wear, noise and increase efficiency by reducing the power wasting in overcoming friction, or to make the mechanism work at all. An engine requires lubrication in at least:

- Between pistons and cylinders
- Small bearings
- Big end bearings

- Main bearings
- Valve gear (The following elements may not be present):
  - o Tappets
  - o Rocker arms
  - o Pushrods
  - o Timing chain or gears. Toothed belts do not require lubrication.

In 2-stroke crankcase scavenged engines, the interior of the crankcase, and therefore the crankshaft, connecting rod and bottom of the pistons are sprayed by the 2-stroke oil in the air-fuel-oil mixture which is then burned along with the fuel. The valve train may be contained in a compartment flooded with lubricant so that no oil pump is required.

In a *splash lubrication system* no oil pump is used. Instead the crankshaft dips into the oil in the sump and due to its high speed, it splashes the crankshaft, connecting rods and bottom of the pistons. The connecting rod big end caps may have an attached scoop to enhance this effect. The valve train may also be sealed in a flooded compartment, or open to the crankshaft in a way that it receives splashed oil and allows it to drain back to the sump. Splash lubrication is common for small 4-stroke engines.

In a *forced* (also called *pressurized*) *lubrication system*, lubrication is accomplished in a closed loop which carries motor oil to the surfaces serviced by the system and then returns the oil to a reservoir. The auxiliary equipment of an engine is typically not serviced by this loop; for instance, an alternator may use ball bearings sealed with its lubricant. The reservoir for the oil is usually the sump, and when this is the case, it is called a *wet sump* system. When there is a different oil reservoir the crankcase still catches it, but it is continuously drained by a dedicated pump; this is called a *dry sump* system.

On its bottom, the sump contains an oil intake covered by a mesh filter which is connected to an oil pump then to an oil filter outside the crankcase, from there it is diverted to the crankshaft main bearings and valve train. The crankcase contains at least one *oil gallery* (a conduit inside a crankcase wall) to which oil is introduced from the oil filter. The main bearings contain a groove through all or half its circumference; the oil enters to these grooves from channels connected to the oil gallery. The crankshaft has drillings which take oil from these grooves and deliver it to the big end bearings. All big end bearings are lubricated this way. A single main bearing may provide oil for 0, 1 or 2 big end bearings. A similar system may be used to lubricate the piston, its gudgeon pin and the small end of its connecting rod; in this system, the connecting rod big end has a groove around the crankshaft and a drilling connected to the groove which distributes oil from there to the bottom of the piston and from then to the cylinder.

Other systems are also used to lubricate the cylinder and piston. Every crankshaft big end may have a nozzle to throw an oil jet to the cylinder and bottom of the piston. That nozzle is in movement relative to the cylinder it lubricates, but always pointed towards it or the corresponding piston. Instead, the nozzle can also be placed fixed in the crankshaft and pointing upwards. Typically a forced lubrication systems have a lubricant flow higher than what is required to lubricate satisfactorily, in order to assist with cooling. Specifically, the lubricant system helps to move heat from the hot engine parts to the cooling liquid (in water cooled engines) or fins (in air cooled engines) which then transfer it to the environment. The lubricant must be designed to be chemically stable and maintain suitable viscosities within the temperature range it encounters in the engine.

Cylinder Configuration

Common cylinder configurations include the straight or inline configuration, the more compact V configuration, and the wider but smoother flat or boxer configuration. Aircraft engines can also adopt a radial configuration, which allows more effective cooling. More unusual configurations such as the H, U, X, and W have also been used.

Multiple cylinder engines have their valve train and crankshaft configured so that pistons are at different parts of their cycle. It is desirable to have the piston's cycles uniformly spaced (this is called *even firing*) especially in forced induction engines; this reduces torque pulsations and makes inline engines with more than 3 cylinders statically balanced in its primary forces. However, some engine configurations require odd firing to achieve better balance that what is possible with even firing. For instance, a 4 stroke I2 engine has better balance when the angle between the crankpins is 180° because the pistons move in opposite directions and inertial forces partially cancel, but this gives an odd firing pattern where one cylinder fires 180° of crankshaft rotation after the other, then no cylinder fires for 540°. With an even firing pattern the pistons would move in unison and the associated forces would add.

Multiple crankshaft configurations do not necessarily need a cylinder head at all because they can instead have a piston at each end of the cylinder called an opposed piston design. Because fuel inlets and outlets are positioned at opposed ends of the cylinder, one can achieve uniflow scavenging, which, as in the four-stroke engine is efficient over a wide range of engine speeds. Thermal efficiency is improved because of a lack of cylinder heads. This design was used in the Junkers Jumo 205 diesel aircraft engine, using two crankshafts at either end of a single bank of cylinders, and most remarkably in the Napier Deltic diesel engines. These used three crankshafts to serve three banks of double-ended cylinders arranged in an equilateral triangle with the crankshafts at the corners. It was also used in single-bank locomotive engines, and is still used in marine propulsion engines and marine auxiliary generators.

Diesel Cycle

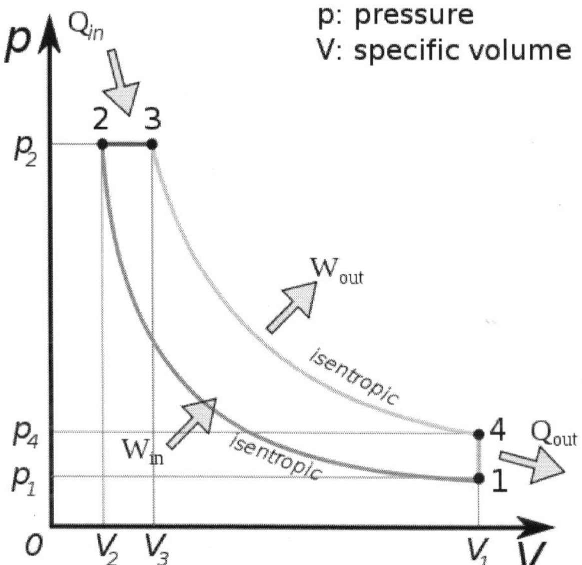

***Figure:** P-v Diagram for the Ideal Diesel cycle. The cycle follows the numbers 1–4 in clockwise direction.*

Most truck and automotive diesel engines use a cycle reminiscent of a four-stroke cycle, but with a compression heating ignition system, rather than needing a separate ignition system. This variation is called the diesel cycle. In the diesel cycle, diesel fuel is injected directly into the cylinder so that combustion occurs at constant pressure, as the piston moves.

Otto cycle: Otto cycle is the typical cycle for most of the cars internal combustion engines, that work using gasoline as a fuel. Otto cycle is exactly the same one that was described for the four-stroke engine. It consists of the same four major steps: Intake, compression, ignition and exhaust.

PV diagram for Otto cycle On the PV-diagram, 1–2: Intake: suction stroke 2–3: Isentropic Compression stroke 3–4: Heat addition stroke 4–5: Exhaust stroke (Isentropic expansion) 5–2: Heat rejection The distance between points 1–2 is the stroke of the engine. By dividing V2/V1, we get: r, where r is called the compression ratio of the engine.

Five-Stroke Engine

In 1879, Nikolaus Otto manufactured and sold double expansion engine (the double and triple expansion principles had ample usage in steam engines), with two small cylinders at both sides of a low-pressure larger cylinder, where a second expansion of exhaust stroke gas took place; the owner returned it, alleging poor performance. In 1906, the concept was incorporated in a car built by EHV (Eisenhuth Horseless Vehicle Company) CT, USA and in the 21st century, ILMOR designed and successfully tested a 5-stroke double expansion internal combustion engine, with high power output and low SFC (Specific Fuel Consumption).

Six-Stroke Engine

The six-stroke engine was invented in 1883. Four kinds of six-stroke use a regular piston in a regular cylinder (Griffin six-stroke, Bajulaz six-stroke, Velozeta six-stroke and Crower six stroke), firing every three crankshaft revolutions. The systems capture the wasted heat of the four-stroke Otto cycle with an injection of air or water.

The Beare Head and "piston charger" engines operate as opposed-piston engines, two pistons in a single cylinder, firing every two revolutions rather more like a regular four-stroke.

Other Cycles

The very first internal combustion engines did not compress the mixture. The first part of the piston downstroke drew in a fuel-air mixture, then the inlet valve closed and, in the remainder of the down-stroke, the fuel-air mixture fired. The exhaust valve opened for the piston upstroke. These attempts at imitating the principle of a steam engine were very inefficient. There are a number of variations of these cycles, most notably the Atkinson and Miller cycles. The diesel cycle is somewhat different.

Split-cycle engines separate the four strokes of intake, compression, combustion and exhaust into two separate but paired cylinders. The first cylinder is used for intake and compression. The compressed air is then transferred through a crossover passage from the compression cylinder into the second cylinder, where combustion and exhaust occur.

A split-cycle engine is really an air compressor on one side with a combustion chamber on the other.

Previous split-cycle engines have had two major problems—poor breathing (volumetric efficiency) and low thermal efficiency. However, new designs are being introduced that seek to address these problems.

The Scuderi Engine addresses the breathing problem by reducing the clearance between the piston and the cylinder head through various turbo charging techniques. The Scuderi design requires the use of outwardly opening valves that enable the piston to move very close to the cylinder head without the interference of the valves. Scuderi addresses the low thermal efficiency via firing ATDC.

Firing ATDC can be accomplished by using high-pressure air in the transfer passage to create sonic flow and high turbulence in the power cylinder.

Combustion Turbines

Gas Turbines

A gas turbine is called a gas turbine because it compresses a gas, usually air. There are three stages to a turbine: 1) air is drawn through a compressor where the temperature rises due to compression, 2) fuel is added in the combuster, and 3) hot air is exhausted through turbines blades which rotate a shaft connected to the compressor.

A gas turbine is a rotary machine similar in principle to a steam turbine and it consists of three main components: a compressor, a combustion chamber, and a turbine. The air, after being compressed in the compressor, is heated by burning fuel in it. About T! of the heated air, combined with the products of combustion, expands in a turbine, producing work output that drives the compressor. The rest (about S!) is available as useful work output.

Jet Engine

The jet engine takes a large volume of hot gas from a combustion process (typically a gas turbine, but rocket forms of jet propulsion often use solid or liquid propellants, and ramjet forms also lack the gas turbine) and feeds it through a nozzle that accelerates the jet to high speed. As the jet accelerates through the nozzle, this creates thrust and in turn does useful work.

Brayton Cycle

A gas turbine is a rotary machine somewhat similar in principle to a steam turbine and it consists of three main components: a compressor, a combustion chamber, and a turbine. The air is compressed by the compressor where a temperature rise occurs, further heated by combustion of injected fuel which heats and expands the air, this energy is tapped by the turbine and exhausted, which powers the compressor and provides thrust.

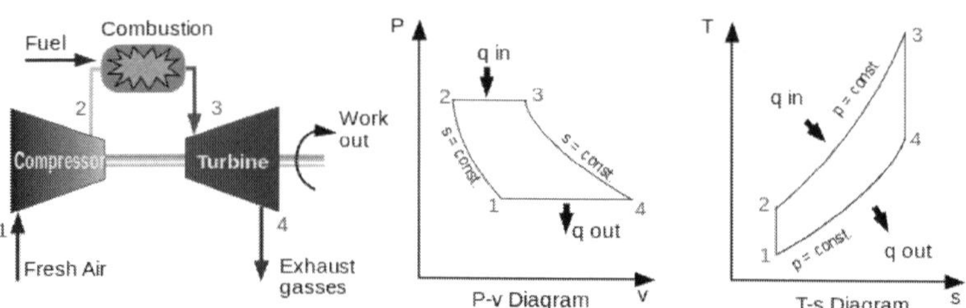

Figure: *Brayton cycle*

Gas turbine cycle engines employ a continuous combustion system where compression, combustion, and expansion occur simultaneously at different places in the engine—giving continuous power. Notably, the combustion takes place at constant pressure, rather than with the Otto cycle, constant volume.

Wankel Engines

The Wankel engine (rotary engine) does not have piston strokes. It operates with the same separation of phases as the four-stroke engine with the phases taking place in separate locations in the engine. In thermodynamic terms it follows the Otto engine cycle, so may be thought of as a "four-phase" engine. While it is true that three power strokes typically occur per rotor revolution, due to the 3:1 revolution ratio of the rotor to the eccentric shaft, only one power stroke per shaft revolution actually occurs. The drive (eccentric) shaft rotates once during every power stroke instead of twice (crankshaft), as in the Otto cycle, giving it a greater power-to-weight ratio than piston engines. This type of engine was most notably used in the Mazda RX-8, the earlier RX-7, and other models.

Forced Induction

Forced induction is the process of delivering compressed air to the intake of an internal combustion engine. A forced induction engine uses a gas compressor to increase the pressure, temperature and density of the air. An engine without forced induction is considered a naturally aspirated engine.

Forced induction is used in the automotive and aviation industry to increase engine power and efficiency. It particularly helps aviation engines, as they need to operate at high altitude.

Forced induction is achieved by a supercharger, where the compressor is directly powered from the engine shaft or, in the turbocharger, from a turbine powered by the engine exhaust.

Fuels and Oxidizers

All internal combustion engines depend on combustion of a chemical fuel, typically with oxygen from the air (though it is possible to inject nitrous oxide to do more of the same thing and gain a power boost). The combustion process typically results in the production of a great quantity of heat, as well as the production of steam and carbon dioxide and other chemicals at very high temperature; the temperature reached is determined by the chemical make up of the fuel and oxidisers, as well as by the compression and other factors.

Fuels

The most common modern fuels are made up of hydrocarbons and are derived mostly from fossil fuels (petroleum). Fossil fuels include diesel fuel, gasoline and petroleum gas, and the rarer use of propane. Except for the fuel delivery components, most internal combustion engines that are designed for gasoline use can run on natural gas or liquefied petroleum gases without major modifications.

Large diesels can run with air mixed with gases and a pilot diesel fuel ignition injection. Liquid and gaseous biofuels, such as ethanol and biodiesel (a form of diesel fuel that is produced from crops that yield triglycerides such as soybean oil), can also be used.

Engines with appropriate modifications can also run on hydrogen gas, wood gas, or charcoal gas, as well as from so-called producer gas made from other convenient biomass.

Experiments have also been conducted using powdered solid fuels, such as the magnesium injection cycle.

Presently, fuels used include:

- Petroleum:
    - o Petroleum spirit (North American term: gasoline, British term: petrol)
    - o Petroleum diesel.
    - o Autogas (liquified petroleum gas).
    - o Compressed natural gas.
    - o Jet fuel (aviation fuel)
    - o Residual fuel

- Coal:
  - o Gasoline can be made from carbon (coal) using the Fischer-Tropsch process
  - o Diesel fuel can be made from carbon using the Fischer-Tropsch process
- Biofuels and vegoils:
  - o Peanut oil and other vegoils.
  - o Biofuels:
    - Biobutanol (replaces gasoline).
    - Biodiesel (replaces petrodiesel).
    - Dimethyl Ether (replaces petrodiesel).
    - Bioethanol and Biomethanol (wood alcohol) and other biofuels.
    - Biogas
- Hydrogen (mainly spacecraft rocket engines)

Even fluidized metal powders and explosives have seen some use. Engines that use gases for fuel are called gas engines and those that use liquid hydrocarbons are called oil engines; however, gasoline engines are also often colloquially referred to as, "gas engines" ("petrol engines" outside North America).

The main limitations on fuels are that it must be easily transportable through the fuel system to the combustion chamber, and that the fuel releases sufficient energy in the form of heat upon combustion to make practical use of the engine.

Diesel engines are generally heavier, noisier, and more powerful at lower speeds than gasoline engines. They are also more fuel-efficient in most circumstances and are used in heavy road vehicles, some automobiles (increasingly so for their increased fuel efficiency over gasoline engines), ships, railway locomotives, and light aircraft. Gasoline engines are used in most other road vehicles including most cars, motorcycles, and mopeds. Note that in Europe, sophisticated diesel-engined cars have taken over about 45% of the market since the 1990s. There are also engines that run on hydrogen, methanol, ethanol, liquefied petroleum gas (LPG), biodiesel, paraffin and tractor vaporizing oil (TVO).

Hydrogen

Hydrogen could eventually replace conventional fossil fuels in traditional internal combustion engines. Alternatively fuel cell technology may come to deliver its promise and the use of the internal combustion engines could even be phased out.

Although there are multiple ways of producing free hydrogen, those methods require converting combustible molecules into hydrogen or consuming electric energy. Unless that electricity is produced from a renewable source—and is not required for other purposes— hydrogen does not solve any energy crisis. In many situations the disadvantage of hydrogen, relative to carbon fuels, is its storage. Liquid hydrogen has extremely low density (14 times lower than water) and requires extensive insulation—whilst gaseous hydrogen requires heavy tankage. Even when liquefied, hydrogen has a higher specific energy but the volumetric energetic storage is still roughly five times lower than gasoline. However, the energy density of hydrogen is considerably higher than that of electric batteries, making it a serious contender as an energy carrier to replace fossil fuels. The 'Hydrogen on Demand' process creates hydrogen as needed, but has other issues, such as the high price of the sodium borohydride that is the raw material.

Oxidizers

Since air is plentiful at the surface of the earth, the oxidizer is typically atmospheric oxygen, which has the advantage of not being stored within the vehicle. This increases the power-to-weight and power-to-volume ratios. Other materials are used for special purposes, often to increase power output or to allow operation under water or in space.

- Compressed air has been commonly used in torpedoes.
- Compressed oxygen, as well as some compressed air, was used in the Japanese Type 93 torpedo. Some submarines carry pure oxygen. Rockets very often use liquid oxygen.
- Nitromethane is added to some racing and model fuels to increase power and control combustion.
- Nitrous oxide has been used—with extra gasoline—in tactical aircraft, and in specially equipped cars to allow short bursts of added power from engines that otherwise run on gasoline and air. It is also used in the Burt Rutan rocket spacecraft.
- Hydrogen peroxide power was under development for German World War II submarines. It may have been used in some non-nuclear submarines, and was used on some rocket engines (notably the Black Arrow and the Me-163 rocket plane).
- Other chemicals such as chlorine or fluorine have been used experimentally, but have not been found practical.

Cooling

Cooling is required to remove excessive heat — over heating can cause engine failure, usually from wear, cracking or warping. Two most common forms of engine cooling are air-cooled and water cooled. Most modern automotive engines and larger engines are water cooled, while most power tool engines and other small engines are air cooled. Some engines (air or water cooled) also have an oil cooler. In some engines, especially for turbine engine blade cooling and liquid rocket engine cooling, fuel is used as a coolant, simultaneously preheating the fuel, before injecting it into a combustion chamber.

Starting

An internal combustion engine is not usually self-starting so an auxiliary machine is required to start it. Many different systems have been used in the past but modern engines are usually started by an electric motor in the small and medium sizes or by compressed air in the large sizes.

Measures of Engine Performance

Engine types vary greatly in a number of different ways:

- energy efficiency
- fuel/propellant consumption (brake specific fuel consumption for shaft engines, thrust specific fuel consumption for jet engines)
- power-to-weight ratio
- thrust to weight ratio
- Torque curves (for shaft engines) thrust lapse (jet engines)
- Compression ratio for piston engines, overall pressure ratio for jet engines and gas turbines

Energy Efficiency

Once ignited and burnt, the combustion products—hot gases—have more available thermal energy than the original compressed fuel-air mixture (which had higher chemical energy). The available energy is manifested as high temperature and pressure that can be translated into work by the engine. In a reciprocating engine, the high-pressure gases inside the cylinders drive the engine's pistons.

Once the available energy has been removed, the remaining hot gases are vented (often by opening a valve or exposing the exhaust outlet) and this allows the piston to return to its previous position (top dead centre, or TDC). The piston can then proceed to the next phase of its cycle, which varies between engines. Any heat that is not translated into work is normally considered a waste product and is removed from the engine either by an air or liquid cooling system.

Internal combustion engines are primarily heat engines, and as such their theoretical efficiency can be calculated by idealized thermodynamic cycles. The efficiency of a theoretical cycle cannot exceed that of the Carnot cycle, whose efficiency is determined by the difference between the lower and upper operating temperatures of the engine. The upper operating temperature of a terrestrial engine is limited by the thermal stability of the materials used to construct it. All metals and alloys eventually melt or decompose, and there is significant researching into ceramic materials that can be made with greater thermal stability and desirable structural properties. Higher thermal stability allows for greater temperature difference between the lower and upper operating temperatures, hence greater thermodynamic efficiency.

The thermodynamic limits assume that the engine is operating under ideal conditions: a frictionless world, ideal gases, perfect insulators, and operation for infinite time. Real world applications introduce complexities that reduce efficiency. For example, a real engine runs best at a specific load, termed its power band. The engine in a car cruising on a highway is usually operating significantly below its ideal load, because it is designed for the higher loads required for rapid acceleration. In addition, factors such as wind resistance reduce overall system efficiency. Engine fuel economy is measured in miles per gallon or in liters per 100 kilometres. The volume of hydrocarbon assumes a standard energy content.

Most steel engines have a thermodynamic limit of 37%. Even when aided with turbochargers and stock efficiency aids, most engines retain an *average* efficiency of about 18%-20%. Rocket engine efficiencies are much better, up to 70%, because they operate at very high temperatures and pressures and can have very high expansion ratios. Electric motors are better still, at around 85-90% efficiency or more, but they rely on an external power source (often another heat engine at a power plant subject to similar thermodynamic efficiency limits).

There are many inventions aimed at increasing the efficiency of IC engines. In general, practical engines are always compromised by trade-offs between different properties such as efficiency, weight, power, heat, response, exhaust emissions, or noise. Sometimes economy also plays a role in not only the cost of manufacturing the engine itself, but also manufacturing and distributing the fuel. Increasing the engine's efficiency brings better fuel economy but only if the fuel cost per energy content is the same.

Measures of Fuel Efficiency and Propellant Efficiency

For stationary and shaft engines including propeller engines, fuel consumption is measured by calculating the brake specific fuel consumption, which measures the mass flow rate of fuel consumption divided by the power produced.

For internal combustion engines in the form of jet engines, the power output varies drastically with airspeed and a less variable measure is used: thrust specific fuel consumption (TSFC), which is the mass of propellant needed to generate impulses that is measured in either pound force-hour or the grams of propellant needed to generate an impulse that measures one kilonewton-second.

For rockets, TSFC can be used, but typically other equivalent measures are traditionally used, such as specific impulse and effective exhaust velocity.

Air and Noise Pollution

Air Pollution

Internal combustion engines such as reciprocating internal combustion engines produce air pollution emissions, due to incomplete combustion of carbonaceous fuel. The main derivatives of the process are carbon dioxide CO_2, water and some soot — also called particulate matter (PM). The effects of inhaling particulate matter have been studied in humans and animals and include asthma, lung cancer, cardiovascular issues, and premature death. There are, however, some additional products of the combustion process that include nitrogen oxides and sulfur and some uncombusted hydrocarbons, depending on the operating conditions and the fuel-air ratio.

Not all of the fuel is completely consumed by the combustion process; a small amount of fuel is present after combustion, and some of it reacts to form oxygenates, such as formaldehyde or acetaldehyde, or hydrocarbons not originally present in the input fuel mixture. Incomplete combustion usually results from insufficient oxygen to achieve the perfect stoichiometric ratio. The flame is "quenched" by the relatively cool cylinder walls, leaving behind unreacted fuel that is expelled with the exhaust. When running at lower speeds, quenching is commonly observed in diesel (compression ignition) engines that run on natural gas. Quenching reduces efficiency and increases knocking, sometimes causing the engine to stall. Incomplete combustion also leads to the production of carbon monoxide (CO).

Further chemicals released are benzene and 1,3-butadiene that are also hazardous air pollutants.

Increasing the amount of air in the engine reduces emissions of incomplete combustion products, but also promotes reaction between oxygen and nitrogen in the air to produce nitrogen oxides (NO_x). NO_x is hazardous to both plant and animal health, and leads to the production of ozone (O_3).

Ozone is not emitted directly; rather, it is a secondary air pollutant, produced in the atmosphere by the reaction of $NO_{"x"}$ and volatile organic compounds in the presence of sunlight. Ground-level ozone is harmful to human health and the environment. Though the same chemical substance, ground-level ozone should not be confused with stratospheric ozone, or the ozone layer, which protects the earth from harmful ultraviolet rays.

Carbon fuels contain sulfur and impurities that eventually produce sulfur monoxides (SO) and sulfur dioxide (SO_2) in the exhaust, which promotes acid rain.

In the United States, nitrogen oxides, PM, carbon monoxide, sulphur dioxide, and ozone, are regulated as criteria air pollutants under the Clean Air Act to levels where human health and welfare are protected. Other pollutants, such as benzene and 1,3-butadiene, are regulated as hazardous air pollutants whose emissions must be lowered as much as possible depending on technological and practical considerations.

Non-Road Engines

The emission standards used by many countries have special requirements for non-road engines which are used by equipment and vehicles that are not operated on the public roadways. The standards are separated from the road vehicles.

Noise Pollution

Significant contributions to noise pollution are made by internal combustion engines. Automobile and truck traffic operating on highways and street systems produce noise, as do aircraft flights due to jet noise, particularly supersonic-capable aircraft. Rocket engines create the most intense noise.

Idling

Internal combustion engines continue to consume fuel and emit pollutants when idling so it is desirable to keep periods of idling to a minimum. Many bus companies now instruct drivers to switch off the engine when the bus is waiting at a terminal.

In the UK (but only in England), the Road Traffic Vehicle Emissions Fixed Penalty Regulations 2002 (Statutory Instrument 2002 No. 1808) introduced the concept of a "*stationary idling offence*". This means that a driver can be ordered "*by an authorised person ... upon production of evidence of his authorisation, require him to stop the running of the engine of that vehicle*" and a "*person who fails to comply ... shall be guilty of an offence and be liable on summary conviction to a fine not exceeding level 3 on the standard scale*". Only a few local authorities have implemented the regulations, one of them being Oxford City Council.

Chapter 7

Component Parts of Internal Combustion Engines

Internal combustion engines come in a wide variety of types, but have certain family resemblances, and thus share many common types of components.

Combustion Chambers

Internal combustion engines can contain any number of combustion chambers (cylinders), with numbers between one and twelve being common, though as many as 36 (Lycoming R-7755) have been used. Having more cylinders in an engine yields two potential benefits: first, the engine can have a larger displacement with smaller individual reciprocating masses, that is, the mass of each piston can be less thus making a smoother-running engine since the engine tends to vibrate as a result of the pistons moving up and down. Doubling the number of the same size cylinders will double the torque and power. The downside to having more pistons is that the engine will tend to weigh more and generate more internal friction as the greater number of pistons rub against the inside of their cylinders. This tends to decrease fuel efficiency and robs the engine of some of its power. For high-performance gasoline engines using current materials and technology, such as the engines found in modern automobiles, there seems to be a point around 10 or 12 cylinders after which the addition of cylinders becomes an overall detriment to performance and efficiency. Although, exceptions such as the W16 engine from Volkswagen exist.

- Most car engines have four to eight cylinders, with some high-performance cars having ten, 12 — or even 16, and some very small cars and trucks having two or three. In previous years, some quite large cars such as the DKW and Saab 92, had two-cylinder or two-stroke engines.

- Radial aircraft engines had from three to 28 cylinders; examples include the small Kinner B-5 and the large Pratt & Whitney R-4360. Larger examples were built as multiple rows.

  As each row contains an odd number of cylinders, to give an even firing sequence for a four-stroke engine, an even number indicates a two- or four-row engine.

  The largest of these was the Lycoming R-7755 with 36 cylinders (four rows of nine cylinders), but it did not enter production.
- Motorcycles commonly have from one to four cylinders, with a few high-performance models having six; although, some 'novelties' exist with 8, 10, or 12.
- Snowmobiles Usually have one to four cylinders and can be both 2-stroke or 4-stroke, normally in the in-line configuration; however, there are again some novelties that exist with V-4 engines
- Small portable appliances such as chainsaws, generators, and domestic lawn mowers most commonly have one cylinder, but two-cylinder chainsaws exist.
- Large reversible two-cycle marine diesels have a minimum of three to over ten cylinders. Freight diesel locomotives usually have around 12 to 20 cylinders due to space limitations, as larger cylinders take more space (volume) per kwh, due to the limit on average piston speed of less than 30 ft/sec on engines lasting more than 40,000 hours under full power.

Ignition System

The ignition system of an internal combustion engines depends on the type of engine and the fuel used. Petrol engines are typically ignited by a precisely timed spark, and diesel engines by compression heating. Historically, outside flame and hot-tube systems were used.

Spark

The mixture is ignited by an electric spark from a spark plug — the timing of which is very precisely controlled. Almost all gasoline engines are of this type.

Diesel engines timing is precisely controlled by the pressure pump and injector.

Compression

Ignition occurs as the temperature of the fuel/air mixture is taken over its autoignition temperature, due to heat generated by the compression of the air during the compression stroke. The vast majority of compression ignition engines are diesels in which the fuel is mixed with the air after the air has reached ignition temperature. In this case, the timing comes from the fuel injection system. Very small model engines for which simplicity and light weight is more important than fuel costs use easily ignited fuels (a mixture of kerosene, ether, and lubricant) and adjustable compression to control ignition timing for starting and running.

Ignition Timing

For reciprocating engines, the point in the cycle at which the fuel-oxidizer mixture is ignited has a direct effect on the efficiency and output of the ICE. The thermodynamics of the idealized Carnot heat engine tells us that an ICE is most efficient if most of the burning takes place at a high temperature, resulting from compression — near top dead centre. The speed of the flame front is directly affected by the compression ratio, fuel mixture temperature, and octane rating or cetane number of the fuel. Leaner mixtures and lower mixture pressures burn more slowly requiring more advanced ignition timing. It is important to have combustion spread by a thermal flame front (deflagration), not by a shock wave. Combustion propagation by a shock wave is called detonation and, in engines, is also known as pinging or Engine knocking.

So at least in gasoline-burning engines, ignition timing is largely a compromise between a later "retarded" spark — which gives greater efficiency with high octane fuel — and an earlier "advanced" spark that avoids detonation with the fuel used. For this reason, high-performance diesel automobile proponents, such as Gale Banks, believe that:

There's only so far you can go with an air-throttled engine on 91-octane gasoline. In other words, it is the fuel, gasoline, that has become the limiting factor. ... While turbocharging has been applied to both gasoline and diesel engines, only limited boost can be added to a gasoline engine before the fuel octane level again becomes a problem. With a diesel, boost pressure is essentially unlimited. It is literally possible to run as much boost as the engine will physically stand before breaking apart. Consequently, engine designers have come to realise that diesels are capable of substantially more power and torque than any comparably sized gasoline engine.

Fuel Systems

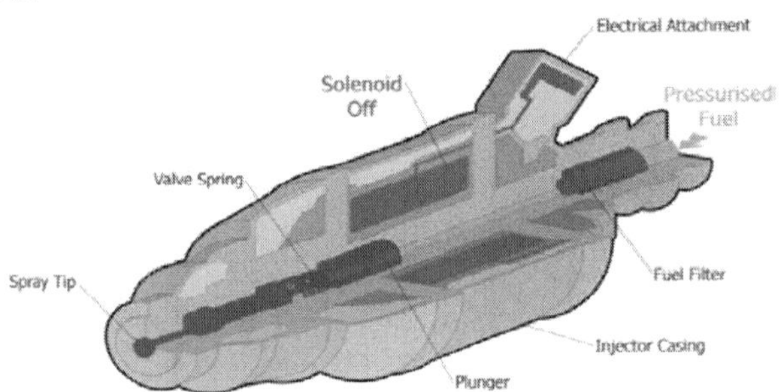

Figure: *Animated cut through diagram of a typical fuel injector, a device used to deliver fuel to the internal combustion engine.*

Fuels burn faster and more efficiently when they present a large surface area to the oxygen in air. Liquid fuels must be atomized to create a fuel-air mixture, traditionally this was done with a carburetor in petrol engines and with fuel injection in diesel engines. Most modern petrol engines now use fuel injection too — though the technology is quite different. While diesel must be injected at an exact point in that engine cycle, no such precision is needed in a petrol engine. However, the lack of lubricity in petrol means that the injectors themselves must be more sophisticated.

Carburetor

Simpler reciprocating engines continue to use a carburetor to supply fuel into the cylinder. Although carburetor technology in automobiles reached a very high degree of sophistication and precision, from the mid-1980s it lost out on cost and flexibility to fuel injection. Simple forms of carburetor remain in widespread use in small engines such as lawn mowers and more sophisticated forms are still used in small motorcycles.

Fuel Injection

Larger gasoline engines used in automobiles have mostly moved to fuel injection systems. Diesel engines have always used fuel injection system because the timing of the injection initiates and controls the combustion.

Autogas engines use either fuel injection systems or open- or closed-loop carburetors.

Fuel Pump

Most internal combustion engines now require a fuel pump. Diesel engines use an all-mechanical precision pump system that delivers a timed injection direct into the combustion chamber, hence requiring a high delivery pressure to overcome the pressure of the combustion chamber. Petrol fuel injection delivers into the inlet tract at atmospheric pressure (or below) and timing is not involved, these pumps are normally driven electrically. Gas turbine and rocket engines use electrical systems.

Other

Other internal combustion engines like jet engines and rocket engines employ various methods of fuel delivery including impinging jets, gas/liquid shear, preburners and others.

Oxidiser-Air Inlet System

Some engines such as solid rockets have oxidisers already within the combustion chamber but in most cases for combustion to occur, a continuous supply of oxidiser must be supplied to the combustion chamber.

Naturally Aspirated Engines

When air is used with piston engines it can simply suck it in as the piston increases the volume of the chamber. However, this gives a maximum of 1 atmosphere of pressure difference across the inlet valves, and at high engine speeds the resulting airflow can limit potential output

Superchargers and Turbochargers

A supercharger is a "forced induction" system which uses a compressor powered by the shaft of the engine which forces air through the valves of the engine to achieve higher flow. When these systems are employed the maximum absolute pressure at the inlet valve is typically around 2 times atmospheric pressure or more.

Turbochargers are another type of forced induction system which has its compressor powered by a gas turbine running off the exhaust gases from the engine.

Turbochargers and superchargers are particularly useful at high altitudes and they are frequently used in aircraft engines.

Duct jet engines use the same basic system, but eschew the piston engine, and replace it with a burner instead.

Liquids

In liquid rocket engines, the oxidiser comes in the form of a liquid and needs to be delivered at high pressure (typically 10-230 bar or 1–

23 MPa) to the combustion chamber. This is normally achieved by the use of a centrifugal pump powered by a gas turbine — a configuration known as a *turbopump*, but it can also be pressure fed.

Parts

For a four-stroke engine, key parts of the engine include the crankshaft (purple), connecting rod (orange), one or more camshafts (red and blue), and valves. For a two-stroke engine, there may simply be an exhaust outlet and fuel inlet instead of a valve system. In both types of engines there are one or more cylinders (grey and green), and for each cylinder there is a spark plug (darker-grey, gasoline engines only), a piston (yellow), and a crankpin (purple). A single sweep of the cylinder by the piston in an upward or downward motion is known as a stroke. The downward stroke that occurs directly after the air-fuel mix passes from the carburetor or fuel injector to the cylinder (where it is ignited) is also known as a power stroke.

A Wankel engine has a triangular rotor that orbits in an epitrochoidal chamber around an eccentric shaft. The four phases of operation (intake, compression, power, and exhaust) take place in what is effectively a moving, variable-volume chamber.

Valves

All four-stroke internal combustion engines employ valves to control the admittance of fuel and air into the combustion chamber. Two-stroke engines use ports in the cylinder bore, covered and uncovered by the piston, though there have been variations such as exhaust valves.

Piston Engine Valves: In piston engines, the valves are grouped into 'inlet valves' which admit the entrance of fuel and air and 'outlet valves' which allow the exhaust gases to escape. Each valve opens once per cycle and the ones that are subject to extreme accelerations are held closed by springs that are typically opened by rods running on a camshaft rotating with the engines' crankshaft.

Control Valves: Continuous combustion engines—as well as piston engines—usually have valves that open and close to admit the fuel and/or air at the startup and shutdown. Some valves feather to adjust the flow to control power or engine speed as well.

Exhaust Systems

Internal combustion engines have to effectively manage the exhaust of the cooled combustion gas from the engine. The exhaust system frequently contains devices to control pollution, both chemical and noise pollution. In addition, for cyclic combustion engines the exhaust system is frequently tuned to improve emptying of the combustion chamber. The majority of exhausts also have systems to prevent heat from reaching places which would encounter damage from it such as heat-sensitive components, often referred to as Exhaust Heat Management.

For jet propulsion internal combustion engines, the 'exhaust system' takes the form of a high velocity nozzle, which generates thrust for the engine and forms a colimated jet of gas that gives the engine its name.

Cooling Systems

Combustion generates a great deal of heat, and some of this transfers to the walls of the engine. Failure will occur if the body of the engine is allowed to reach too high a temperature; either the engine will physically fail, or any lubricants used will degrade to the point that they no longer protect the engine. The lubricants must be clean as dirty lubricants may lead to over formation of sludge in the engines.

Cooling systems usually employ air (air-cooled) or liquid (usually water) cooling, while some very hot engines using radiative cooling (especially some rocket engines). Some high-altitude rocket engines use ablative cooling, where the walls gradually erode in a controlled fashion. Rockets in particular can use regenerative cooling, which uses the fuel to cool the solid parts of the engine.

Piston: A piston is a component of reciprocating engines. It is located in a cylinder and is made gas-tight by piston rings. Its purpose is to transfer force from expanding gas in the cylinder to the crankshaft via a piston rod and/or connecting rod. In two-stroke engines the piston also acts as a valve by covering and uncovering ports in the cylinder wall.

Propelling Nozzle: For jet engine forms of internal combustion engines, a propelling nozzle is present. This takes the high temperature, high pressure exhaust and expands and cools it. The exhaust leaves the nozzle going at much higher speed and provides thrust, as well as constricting the flow from the engine and raising the pressure in the rest of the engine, giving greater thrust for the exhaust mass that exits.

Crankshaft: Most reciprocating internal combustion engines end up turning a shaft. This means that the linear motion of a piston must be converted into rotation. This is typically achieved by a crankshaft.

Flywheels: The flywheel is a disk or wheel attached to the crank, forming an inertial mass that stores rotational energy. In engines with only a single cylinder the flywheel is essential to carry energy over from the power stroke into a subsequent compression stroke. Flywheels are present in most reciprocating engines to smooth out the power delivery over each rotation of the crank and in most automotive engines also mount a gear ring for a starter. The rotational inertia of the flywheel also allows a much slower minimum unloaded speed and also improves the smoothness at idle. The flywheel may also perform a part of the balancing of the system and so by itself be out of balance, although most engines will use a neutral balance for the flywheel, enabling it to be balanced in a separate operation. The flywheel is also used as a mounting for the clutch or a torque converter in most automotive applications.

Starter Systems

All internal combustion engines require some form of system to get them into operation. Most piston engines use a starter motor powered by the same battery as runs the rest of the electric systems. Large jet engines and gas turbines are started with a compressed air motor that is geared to one of the engine's driveshafts. Compressed air can be supplied from another engine, a unit on the ground or by the aircraft's APU. Small internal combustion engines are often started by pull cords. Motorcycles of all sizes were traditionally kick-started, though all but the smallest are now electric-start. Large stationary and marine engines may be started by the timed injection of compressed air into the cylinders — or occasionally with cartridges. Jump starting refers to assistance from another battery (typically when the fitted battery is discharged), while bump starting refers to an alternative method of starting by the application of some external force, e.g. rolling down a hill.

Heat Shielding Systems

These systems often work in combination with engine cooling and exhaust systems. Heat shielding is necessary to prevent engine heat from damaging heat-sensitive components. The majority of older cars use simple steel heat shielding to reduce thermal radiation and convection. It is now most common for modern cars are to use aluminium heat shielding which has a lower density, can be easily formed and does not corrode in the same way as steel. Higher performance vehicles are beginning to use ceramic heat shielding as this can withstand far higher temperatures as well as further reductions in heat transfer.

Lubrication Systems

Internal combustions engines require lubrication in operation that moving parts slide smoothly over each other. Insufficient lubrication subjects the parts of the engine to metal-to-metal contact, friction, heat build-up, rapid wear often culminating in parts becoming friction welded together e.g. pistons in their cylinders. Big end bearings seizing up will sometimes lead to a connecting rod breaking and poking out through the crankcase.

Several different types of lubrication systems are used. Simple two-stroke engines are lubricated by oil mixed into the fuel or injected into the induction stream as a spray. Early slow-speed stationary and marine engines were lubricated by gravity from small chambers similar to those used on steam engines at the time — with an engine tender refilling these as needed. As engines were adapted for automotive and aircraft use, the need for a high power-to-weight ratio led to increased speeds, higher temperatures, and greater pressure on bearings which in turn required pressure-lubrication for crank bearings and connecting-rod journals. This was provided either by a direct lubrication from a pump, or indirectly by a jet of oil directed at pickup cups on the connecting rod ends which had the advantage of providing higher pressures as the engine speed increased.

Control Systems

Most engines require one or more systems to start and shut down the engine and to control parameters such as the power, speed, torque, pollution, combustion temperature, and efficiency and to stabilise the engine from modes of operation that may induce self-damage such as pre-ignition. Such systems may be referred to as engine control units.

Many control systems today are digital, and are frequently termed FADEC (Full Authority Digital Electronic Control) systems.

Diagnostic Systems

Engine On Board Diagnostics (also known as OBD) is a computerized system that allows for electronic diagnosis of a vehicles' powerplant. The first generation, known as *OBD1*, was introduced 10 years after the U.S. Congress passed the Clean Air Act in 1970 as a way to monitor a vehicles' fuel injection system. *OBD2*, the second generation of computerized on-board diagnostics, was codified and recommended by the California Air Resource Board in 1994 and became mandatory equipment aboard all vehicles sold in the United States as of 1996.

Engine Efficiency

Engine efficiency of thermal engines is the relationship between the total energy contained in the fuel, and the amount of energy used to perform useful work. There are two classifications of thermal engines-

1. Internal combustion (gasoline, diesel and gas turbine, i.e., Brayton cycle engines) and
2. External combustion engines (steam piston, steam turbine, and the Stirling cycle engine).

Each of these engines has thermal efficiency characteristics that are unique to it.

Mathematical Definition

The efficiency of engine is defined as ratio of the useful work done to the heat provided.

$$\eta = \frac{work\ done}{heat\ absorbed} = \frac{Q1 - Q2}{Q1}$$

where, $Q1$ is the heat absorbed and $Q1 - Q2$ is the work done.

Please note that the term work done relates to the power delivered at the clutch or at the driveshaft.

This means the friction and other losses are subtracted from the work done by thermodynamic expansion. Thus an engine not delivering any work to the outside environment has zero efficiency.

Compression Ratio

The efficiency of internal combustion engines depends on several factors, one of which is the compression ratio. Most gasoline (petrol) engines have a *geometric* compression ratio (the compression ratio calculated purely from the geometry of the mechanical parts) of 10:1 (premium fuel) or 9:1 (regular fuel), with some engines reaching a ratio of 12:1 or more. The greater the compression ratio the more efficient is the engine, in principle, and higher compression-ratio conventional engines in principle need gasoline with higher octane value, though this simplistic analysis is complicated by the difference between actual and geometric compression ratios.

High octane value inhibits the fuel's tendency to burn nearly instantaneously (known as *detonation* or *knock*) at high compression/ high heat conditions. However, in engines that utilise compression rather than spark ignition, by means of very high compression ratios (14-25:1), such as the diesel engine or Bourke engine, high octane fuel is not necessary. In fact, lower-octane fuels, typically rated by cetane number, are preferable in these applications because they are more easily ignited under compression.

Under part throttle conditions (i.e. when the throttle is less than fully open), the *effective* compression ratio is less than when the engine is operating at full throttle, due to the simple fact that the incoming fuel-air mixture is being restricted and cannot fill the chamber to full atmospheric pressure. The engine efficiency is less than when the engine is operating at full throttle.

One solution to this fact is to shift the load in a multi-cylinder engine from some of the cylinders (by deactivating them) to the remaining cylinders so that they may operate under higher individual loads and with correspondingly higher effective compression ratios. This technique is known as variable displacement.

Diesel engines have a compression ratio between 14:1 to 25:1. In this case the general rule does not apply because Diesels with compression ratios over 20:1 are indirect injection diesels. These use a prechamber to make possible high RPM operation as is required in automobiles and light trucks. The thermal and gas dynamic losses from the prechamber result in direct injection Diesels (despite their lower compression ratio) being more efficient.

Friction

An engine has many moving parts that produce friction. Some of these friction forces remain constant (as long as applied load is constant); some of these friction losses increase as engine speed increases, such as piston side forces and connecting bearing forces (due to increased inertia forces from the oscillating piston).

A few friction forces decrease at higher speed, such as the friction force on the cam's lobes used to operate the inlet and outlet valves (the valves' inertia at high speed tends to pull the cam follower away from the cam lobe). Along with friction forces, an operating engine has *pumping losses*, which is the work required to move air into and out of the cylinders. This pumping loss is minimal at low speed, but increases approximately as the square of the speed, until at rated power an engine is using about 20% of total power production to overcome friction and pumping losses.

Oxygen

Air is approximately 21% oxygen. If there is not enough oxygen for proper combustion, the fuel will not burn completely and will produce less energy. An excessively rich air fuel ratio will increase pollutants from the engine. If all of the oxygen is consumed because there is too much fuel, engine's power is reduced.

There are a few exceptions where introducing fuel upstream of the combustion chamber can cool down the incoming air through evaporative cooling. The extra fuel that is not burned in the combustion chamber cools down the intake air resulting in more power. With direct injection this effect is not as dramatic but it can cool down the combustion chamber enough to reduce certain pollutants such as nitrogen oxides (NOx), while raising others such as partially decomposed hydrocarbons.

The air-fuel mix is drawn into an engine because downward motion of the pistons induces a partial vacuum. A compressor can additionally be used to force a larger charge (forced induction) into the cylinder to produce more power. The compressor is either mechanically driven supercharging or exhaust driven turbocharging. Either way, forced induction increases the air pressure exterior to the cylinder inlet port.

There are other methods to increase the amount of oxygen available inside the engine; one of them, is to inject nitrous oxide, (N2O) to the mixture, and some engines use nitromethane, a fuel that provides the oxygen itself it needs to burn. Because of that, the mixture could be 1 part of fuel and 3 parts of air; thus, it is possible to burn more fuel inside the engine, and get higher power outputs...

Internal Combustion Engines

Gasoline (Petrol) Engines

Modern gasoline engines have a maximum thermal efficiency of about 25% to 30% when used to power a car. In other words, even when the engine is operating at its point of maximum thermal efficiency, of the total heat energy released by the gasoline consumed, about 70-75% is rejected as heat without being turned into useful work, i.e. turning the crankshaft. Approximately half of this rejected heat is carried away by the exhaust gases, and half passes through the cylinder walls or cylinder head into the engine cooling system, and is passed to the atmosphere via the cooling system radiator. Some of the work generated is also lost as friction, noise, air turbulence, and work used to turn engine equipment and appliances such as water and oil pumps and the electrical generator, leaving only about 25-30% of the energy released by the fuel consumed available to move the vehicle.

At idle, the thermal efficiency is zero, since no usable work is being drawn from the engine. At low speeds, gasoline engines suffer efficiency losses at small throttle openings from the high turbulence and frictional (head) loss when the incoming air must fight its way around the nearly closed throttle; diesel engines do not suffer this loss because the incoming air is not throttled.

At high speeds, efficiency in both types of engine is reduced by pumping and mechanical frictional losses, and the shorter period within which combustion has to take place. Engine efficiency peaks in most applications at around 75% of rated engine power, which is also the range of greatest engine torque (e.g. in most modern passenger automobile engines with a redline of about 6,000 RPM, maximum torque is obtained at about 4,500 RPM, and maximum engine power is obtained at about 6,000 RPM). At all other combinations of engine speed and torque, the thermal efficiency is less than this maximum.

A gasoline engine burns a mix of gasoline and air, consisting of a range of about twelve to eighteen parts (by weight) of air to one part of fuel (by weight). A mixture with a 14.7:1 air/fuel ratio is said to be stoichiometric, that is when burned, 100% of the fuel and the oxygen are consumed. Mixtures with slightly less fuel, called lean burn are more efficient. The combustion is a reaction which uses the air's oxygen content to combine with the fuel, which is a mixture of several hydrocarbons, resulting in water vapor, carbon dioxide, and sometimes carbon monoxide and partially burned hydrocarbons. In addition, at high temperatures the oxygen tends to combine with nitrogen, forming oxides of nitrogen (usually referred to as *NOx*, since the number of oxygen atoms in the compound can vary, thus the "X" subscript). This mixture, along with the unused nitrogen and other trace atmospheric elements, is what we see in the exhaust.

In the past 3–4 years, GDI (Gasoline Direct Injection) increased the efficiency of the engines equipped with this fuelling system up to 35%. Currently, the technology is available in a wide variety of vehicles ranging from less expensive cars produced by Mazda, Ford and Chevrolet to more expensive cars produced by BMW, Mercedes-Benz, and Volkswagen Auto Group.

Diesel Engines

Engines using the Diesel cycle are usually more efficient, although the Diesel cycle itself is less efficient at equal compression ratios. Since diesel engines use much higher compression ratios (the heat of compression is used to ignite the slow-burning diesel fuel), that higher ratio more to air pumping losses within the engine. Modern turbo-diesel engines are using electronically controlled, common-rail fuel injection, that increases the efficiency up to 50% with the help of geometrically variable turbo-charging system; this also increases the engines' torque at low engine speeds (1200-1800RPM).

Gas Turbine

The gas turbine is most efficient at maximum power output in the same way reciprocating engines are most efficient at maximum load. The difference is that at lower rotational speed the pressure of the compressed air drops and thus thermal and fuel efficiency drop dramatically. Efficiency declines steadily with reduced power output and is very poor in the low power range - the same is true in reciprocating engines, the friction losses at 3000 RPM are almost the same whether the engine is under 10% load or not having any useful output on the driveshaft. The inertia of high speed gas turbine together with the low air pressure under low speed cause it to have a significant lag which many drivers are unwilling to cope with.

Today the gas turbine is not used for automobiles and trucks because the usage patterns dictate varying loads, including idling speeds. General Motors at one time manufactured a bus powered by a gas turbine, but due to the economy where crude oil prices rose exponentially (1970's) this concept was abandoned. Rover, Chrysler, and Toyota also built prototypes of turbine powered cars, Chrysler building a short prototype series of them for real-world evaluation. Driving comfort was good, but overall economy lacked due to reasons mentioned above. This is also why gas turbines can be used for permanent and peak power electric plants. In this application they are only run at or close to full power where they are efficient or shut down when not needed.

Gas turbines do have advantage in power density - gas turbines are used as the engines in heavy armored vehicles and armored tanks and in power generators in jet fighters.

One other factor negatively affecting the gas turbine efficiency is the ambient air temperature. With increasing temperature, intake air becomes less dense and therefore the gas turbine experiences power loss proportional to the increase in ambient air temperature.

External Combustion Engines

Steam Engine

Piston Engine: Steam engines and turbines operate on the Rankine cycle which has a maximum Carnot efficiency of 63% for practical engines.

The efficiency of steam engines is primarily related to the steam temperature and pressure and the number of stages or *expansions*. Steam engine efficiency improved as the operating principles were discovered, which lead to the development of the science of thermodynamics.

In earliest steam engines the boiler was considered part of the engine. Today they are considered separate, so it is necessary to know whether stated efficiency is overall, which includes the boiler, or just of the engine.

Comparisons of efficiency and power of the early steam engines is difficult for several reasons: 1) there was no standard weight for a bushel of coal, which could be anywhere from 82 to 96 pounds. 2) There was no standard heating value for coal, and probably no way to measure heating value. The coals had much higher heating value than today's steam coals, with 13,500 BTU/pound sometimes mentioned. 3) Efficiency was reported as "duty", meaning how many foot pounds of work lifting water were produced, but the mechanical pumping efficiency is not known.

The first piston steam engine, developed by Thomas Newcomen around 1710, was slightly over one half percent (0.5%) efficient. It operated with steam at near atmospheric pressure drawn into the cylinder by the load, then condensed by a spray of cold water into the steam filled cylinder, causing a partial vacuum in the cylinder and the pressure of the atmosphere to drive the piston down. Using the cylinder as the vessel in which to condense the steam also cooled the cylinder, so that some of the heat in the incoming steam on the next cycle was lost in warming the cylinder, reducing the thermal efficiency. Improvements made by John Smeaton to the Newcomen engine increased the efficiency to over 1%.

James Watt made several improvements to the Newcomen engine, the most significant of which was the external condenser, which prevented the cooling water from cooling the cylinder. Watt's engine operated with steam at slightly above atmospheric pressure. Watt's improvements increased efficiency by a factor of over 2.5. The lack of general mechanical ability, including skilled mechanics, machine tools, and manufacturing methods, limited the efficiency of actual engines and their design until about 1840.

Higher pressures engines were developed by Oliver Evans and independently by Richard Trevithick. These engines were not very efficient but had high power-to-weight ratio, allowing them to be used for powering locomotives and boats.

The centrifugal governor, which had first been used by Watt to maintain constant speed, worked by throttling the inlet steam, which lowered the pressure, resulting in a loss of efficiency on the high (above atmospheric) pressure engines. Later control methods reduced or eliminated this pressure loss.

The improved valving mechanism of the Corliss steam engine (Ptd. 1849) was better able to adjust speed with varying load and increased efficiency by about 30%. The Corliss engine had separate valves and headers for the inlet and exhaust steam so the hot feed steam never contacted the cooler exhaust pots and valving. The valves were quick acting, which reduced the amount of throttling of the steam and resulted in faster response. Instead of operating a throttling valve, the governor was used to adjust the valve timing to give a variable steam cut off. The variable cut off was responsible for a major portion of the efficiency increase of the Corliss engine.

Others before Corliss had at least part of this idea, including Zachariah Allen, who patented variable cut off, but lack of demand, increased cost and complexity and poorly developed machining technology delayed introduction until Corliss.

The Porter-Allen high speed engine (ca. 1862) operated at from three to five times the speed of other similar sized engines. The higher speed minimized the amount of condensation in the cylinder, resulting in increased efficiency.

Compound engines gave further improvements in efficiency. By the 1870s triple expansion engines were being used on ships. Compound engines allowed ships to carry less coal than freight. Compound engines were used on some locomotives but were not widely adopted because of their mechanical complexity.

The most efficient reciprocating steam engine design (per stage) was the uniflow engine, but by the time it appeared steam was being displaced by diesel engines, which were even more efficient and had the advantage of requiring less labour for coal handling and oil being a more dense fuel displaced less cargo.

Steam Turbine

The steam turbine is the most efficient steam engine and for this reason is universally used for electrical generation. Steam expansion in a turbine is nearly continuous, which makes a turbine comparable to a very large number of expansion stages. Steam fossil fuel power stations operating at the critical point have efficiencies in the low 40% range. Turbines produce direct rotary motion and are far more compact and weigh far less than reciprocating engines and can be controlled to within a very constant speed.

Stirling Engines

The Stirling cycle engine has the highest theoretical efficiency of any thermal engine but it is more expensive to make and is not competitive with other types for normal commercial use.

Horsepower

Horsepower (hp) is a unit of measurement of power (the rate at which work is done). There are many different standards and types of horsepower. The most common horsepower—especially for electrical power—is 1 hp = 746 watts. The term was adopted in the late 18th century by Scottish engineer James Watt to compare the output of steam engines with the power of draft horses. It was later expanded to include the output power of other types of piston engines, as well as turbines, electric motors and other machinery. The definition of the unit varied between geographical regions. Most countries now use the SI unit *watt* for measurement of power. With the implementation of the EU Directive 80/181/EEC on January 1, 2010, the use of horsepower in the EU is permitted only as a supplementary unit.

Definitions of Term

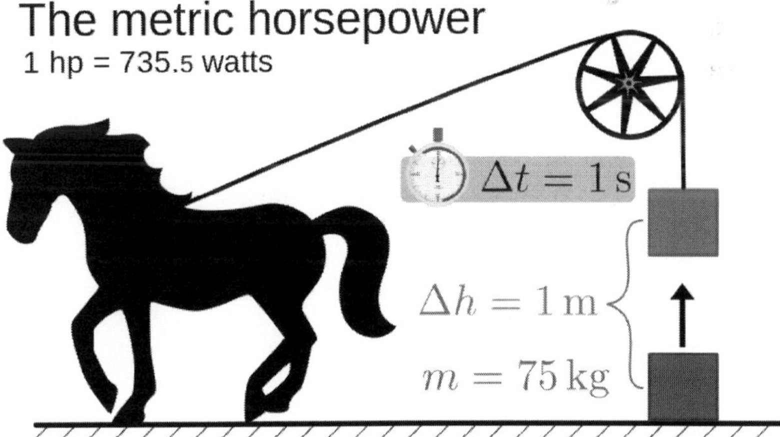

***Figure:** One* metric horsepower *is needed to lift 75 kilograms (avg. body weight of a person) by 1 metre (3.28 feet) in 1 second.*

Units called "horsepower" have differing definitions:

- The *mechanical horsepower*, also known as *imperial horsepower*, of exactly 550 foot-pounds per second is approximately equivalent to 745.7 watts.

- The *metric horsepower* of 75 kgf-m per second is approximately equivalent to 735.5 watts.
    - o The *Pferdestärke* PS (German translation of horsepower) is a name for a group of similar power measurements used in Germany around the end of the 19th century, all of about one metric horsepower in size.
- The *boiler horsepower* is used for rating steam boilers and is equivalent to 34.5 pounds of water evaporated per hour at 212 degrees Fahrenheit, or 9,809.5 watts.
- One horsepower for rating electric motors is equal to 746 watts.
- Continental European electric motors used to have dual ratings, using a conversion rate of 0.735 kW for 1 hp
- British Royal Automobile Club (RAC) horsepower is one of the tax horsepower systems adopted around Europe which make an estimate based on several engine dimensions.

History of the Unit

The development of the steam engine provided a reason to compare the output of horses with that of the engines that could replace them. In 1702, Thomas Savery wrote in The Miner's Friend:

"So that an engine which will raise as much water as two horses, working together at one time in such a work, can do, and for which there must be constantly kept ten or twelve horses for doing the same. Then I say, such an engine may be made large enough to do the work required in employing eight, ten, fifteen, or twenty horses to be constantly maintained and kept for doing such a work..."

The idea was later used by James Watt to help market his improved steam engine. He had previously agreed to take royalties of one third of the savings in coal from the older Newcomen steam engines. This royalty scheme did not work with customers who did not have existing steam engines but used horses instead.

Watt determined that a horse could turn a mill wheel 144 times in an hour (or 2.4 times a minute). The wheel was 12 feet in radius; therefore, the horse travelled $2.4 \times 2\pi \times 12$ feet in one minute. Watt judged that the horse could pull with a force of 180 pounds. So:

$$P = \frac{W}{t} = \frac{Fd}{t} = \frac{(180 \text{ lbf})(2.4 \times 2\pi \times 12 \text{ ft})}{1 \text{ min}} = 32{,}572 \, \frac{\text{ft·lbf}}{\text{min}}.$$

This was rounded to an even 33,000 ft·lbf/min. Watt calculated the power as 33,000 ft-lb per minute.

Watt determined that a pony could lift an average 220 lbf (0.98 kN) 100 ft (30 m) per minute over a four-hour working shift. Watt then judged a horse was 50% more powerful than a pony and thus arrived at the 33,000 ft·lbf/min figure. *Engineering in History* recounts that John Smeaton initially estimated that a horse could produce 22,916 foot-pounds per minute.

John Desaguliers had previously suggested 44,000 foot-pounds per minute and Tredgold 27,500 foot-pounds per minute. "Watt found by experiment in 1782 that a 'brewery horse' could produce 32,400 foot-pounds per minute." James Watt and Matthew Boulton standardized that figure at 33,000 the next year.

Most observers familiar with horses and their capabilities estimate that Watt was either a bit optimistic or intended to underpromise and overdeliver; few horses can maintain that effort for long. Regardless, comparison with a horse proved to be an enduring marketing tool.

In 1993, R. D. Stevenson and R. J. Wassersug published an article calculating the upper limit to an animal's power output. The peak power over a few seconds has been measured to be as high as 14.9 hp. However, Stevenson and Wassersug observe that for sustained activity, a work rate of about 1 hp per horse is consistent with agricultural advice from both 19th and 20th century sources.

When considering human-powered equipment, a healthy human can produce about 1.2 hp briefly and sustain about 0.1 hp indefinitely; trained athletes can manage up to about 2.5 hp briefly and 0.3 hp for a period of several hours.

Calculating Power

If torque and angular speed are known, using a coherent system of units (such as SI), the power may be calculated using the relationship;

$$P = \tau\omega$$

where P is power, τ is torque, and ω is angular speed. When using other units or if the speed is in revolutions per unit time rather than radians, a conversion factor has to be included. When torque is in pound-foot units, rotational speed (f) is in rpm and power is required in horsepower:

$$P(\text{hp}) = \frac{\tau(\text{ft·lbf}) \times f(\text{rpm})}{5252}$$

The constant 5252 is the rounded value of (33,000 ft·lbf/min)/(2π rad/rev).

When torque is in inch pounds:

$$P(\text{hp}) = \frac{\tau(\text{in·lbf}) \times f(\text{rpm})}{63{,}025}$$

The constant 63,025 is the rounded value of (33,000 ft ·lbf/min) × (12 in/ft)/(2 π rad/rev).

Current Definitions

The following definitions have been widely used:

MECHANICAL HORSEPOWER HP(I)	≡ 33,000 FT-LB$_F$/MIN = 550 FT ·LB$_F$/S ≈ 17696 LB$_M$ ·FT^2/S^3 = 745.69 W
Metric horsepower hp(M) - also *PS*, "*cv*, hk, pk, ks *or* ch	≡ 75 kg$_f$ m/s ≡ 735.49875 W
Electrical horsepower hp(E)	≡ 746 W
Boiler horsepower hp(S)	≡ 33,475 BTU/h = 9,812.5 W
Hydraulic horsepower	= flow rate (US gal/min) × pressure (psi) × 7/12,000 or = flow rate (US gal/min) × pressure (psi) / 1714 = 550 ft ·lb$_f$/s = 745.69988145 W

In certain situations it is necessary to distinguish between the various definitions of horsepower and thus a suffix is added: hp(I) for mechanical (or imperial) horsepower, hp(M) for metric horsepower, hp(S) for boiler (or steam) horsepower and hp(E) for electrical horsepower.

Hydraulic horsepower is equivalent to mechanical horsepower. The formula given above is for conversion to mechanical horsepower from the factors acting on a hydraulic system.

Mechanical Horsepower

Assuming the third CGPM (1901, CR 70) definition of standard gravity, g_n=9.80665 m/s^2, is used to define the pound-force as well as the kilogram force, and the international avoirdupois pound (1959), one mechanical horsepower is:

1 HP	≡ 33,000 FT-LBF/MIN	BY DEFINITION		
	= 550 ft·lbf/s	since	1 min	= 60 s
	= 550×0.3048×0.453592376 m·kgf/s	since	1 ft	= 0.3048 m and
	= 76.0402259128 kgf·m/s		1 lb	= 0.453592376 kg
	= 76.0402259128×9.80665 kg·m²/s³		*g*	= 9.80665 m/s²
	= 745.699881448 W	since	1 W	≡ 1 J/s = 1 N·m/s = 1 (kg·m/s²)·(m/s)

Or given that 1 hp = 550 ft·lb_f/s, 1 ft = 0.3048 m, 1 lb_f H≈ 4.448 N, 1 J = 1 N·m, 1 W = 1 J/s: 1 hp H≈ 746 W

Metric Gorsepower (PS, cv, hk, pk, ks, ch)

The various units used to indicate this definition (*PS*, *cv*, *hk*, *pk*, *ks* and *ch*) all translate to *horse power* in English, so it is common to see these values referred to as *horsepower* or *hp* in the press releases or media coverage of the German, French, Italian, and Japanese automobile companies. British manufacturers often intermix metric horsepower and mechanical horsepower depending on the origin of the engine in question. Sometimes the metric horsepower rating of an engine is conservative enough so that the same figure can be used for both 80/1269/EEC with metric hp and SAE J1349 with imperial hp.

DIN 66036 defines one metric horsepower as the power to raise a mass of 75 kilograms against the earth's gravitational force over a distance of one metre in one second; this is equivalent to 735.49875 W or 98.6% of an imperial mechanical horsepower.

In 1972, the PS was rendered obsolete by EEC directives, when it was replaced by the kilowatt as the official power measuring unit. It is still in use for commercial and advertising purposes, in addition to the kW rating, as many customers are still not familiar with the use of kilowatts for engines.

Other names for the metric horsepower are the Dutch *paardenkracht* (pk), the French *cheval* (ch), the Portuguese *cavalo-vapor* (cv), the Russian, the Swedish *hästkraft* (hk), the Finnish *hevosvoima* (hv), the Norwegian and Danish *hestekraft* (hk), the Hungarian *lóerõ* (LE), the Czech *koòská síla* and Slovak *konská sila* (k or ks), the Bosnian/Croatian/Serbian *konjska snaga* (KS), the Bulgarian *Êîíñêà ñèëà*, the Macedonian *Êîœñêà ñèëà* (KC), the Polish *koñ mechaniczny*, Slovenian *konjska moè* (KM) and the Romanian *cal-putere* (CP), which all equal the German *Pferdestärke* (PS).

In the 19th century, the French had their own unit, which they used instead of the CV or horsepower. It was called the poncelet and was abbreviated *p*.

French and Italian Tax Horsepower (CV)

In addition, the capital form *CV* is used in Italy and France as a unit for tax horsepower, short for, respectively, *cavalli vapore* and *chevaux vapeur* (*steam horses*). CV is a non-linear rating of a motor vehicle for tax purposes. The CV rating, or fiscal power, is $\left(\frac{P}{40}\right)^{1.6}+\frac{U}{45}$, where P is the maximum power in kilowatts and U is the amount of carbon dioxide (CO_2) emitted in grams per kilometre. The term for CO_2 measurements has been included in the definition only since 1998, so older ratings in CV are not directly comparable. The fiscal power has found its way into naming of automobile models, such as the popular Citroën deux-chevaux. The *cheval-vapeur* (ch) unit should not be confused with the French *cheval fiscal* (CV).

Electrical Horsepower

The horsepower used for electrical machines is defined as exactly 746 W. The nameplates on electrical motors show their power output, not their power input. Outside the United States watts or kilowatts are generally used for electric motor ratings and in such usage it is the input power that is stated.

Boiler Horsepower

Boiler horsepower is a boiler's capacity to deliver steam to a steam engine and is not the same unit of power as the 550 ft-lb/s definition. One boiler horsepower is equal to the power required to evaporate 34.5 lb of fresh water at 212°F in one hour. In the early days of steam use, the boiler horsepower was roughly comparable to the horsepower of engines fed by the boiler. Boiler horsepower is still used to measure boiler output in industrial boiler engineering in Australia, the US, and New Zealand. Boiler horsepower is abbreviated BHP, not to be confused with brake horsepower, below, which is also called BHP.

Drawbar Horsepower

Drawbar horsepower (dbhp) is the power a railway locomotive has available to haul a train or an agricultural tractor to pull an implement. This is a measured figure rather than a calculated one. A special railway car called a dynamometre car coupled behind the locomotive keeps a continuous record of the drawbar pull exerted, and the speed. From these, the power generated can be calculated. To determine the maximum power available, a controllable load is required; it is normally a second locomotive with its brakes applied, in addition to a static load.

If the drawbar force (F) is measured in pounds-force (lbf) and speed (v) is measured in miles per hour (mph), then the drawbar power (P) in horsepower (hp) is:

$$P/\text{hp} = \frac{(F/\text{lbf})(v/\text{mph})}{375}$$

Example: How much power is needed to pull a drawbar load of 2,025 pounds-force at 5 miles per hour?

$$P/\text{hp} = \frac{2025 \times 5}{375} = 27$$

The constant 375 is because 1 hp = 375 lbf·mph. If other units are used, the constant is different. When using a coherent system of units, such as SI (watts, newtons, and metres per second), no constant is needed, and the formula becomes $P = Fv$. This formula may also be used to calculate the horsepower of a jet engine, using the speed of the jet and the thrust required to maintain that speed.

Example: How much power is generated with a thrust of 4,000 pounds at 400 miles per hour?

$$P/\text{hp} = \frac{4000 \times 400}{375} = 4266.7$$

RAC Horsepower (Taxable Horsepower)

This measure was instituted by the Royal Automobile Club in Britain and was used to denote the power of early 20th-century British cars. Many cars took their names from this figure (hence the Austin Seven and Riley Nine), while others had names such as "40/50 hp", which indicated the RAC figure followed by the true measured power.

Taxable horsepower does not reflect developed horsepower; rather, it is a calculated figure based on the engine's bore size, number of cylinders, and a (now archaic) presumption of engine efficiency. As new engines were designed with ever-increasing efficiency, it was no longer a useful measure, but was kept in use by UK regulations which used the rating for tax purposes.

$$RACh.p. = (D^2 * n)/2.5$$

where

D is the diameter (or bore) of the cylinder in inches

n is the number of cylinders

This is equal to the engine displacement in cubic inches divided by 10π then divided again by the stroke in inches.

Since taxable horsepower was computed based on bore and number of cylinders, not based on actual displacement, it gave rise to engines with 'undersquare' dimensions (bore smaller than stroke) this tended to impose an artificially low limit on rotational speed (rpm), hampering the potential power output and efficiency of the engine.

The situation persisted for several generations of four- and six-cylinder British engines: for example, Jaguar's 3.4-litre XK engine of the 1950s had six cylinders with a bore of 83 mm (3.27 in) and a stroke of 106 mm (4.17 in), where most American automakers had long since moved to oversquare (large bore, short stroke) V-8s (see, for example, the early Chrysler Hemi).

Measurement

The power of an engine may be measured or estimated at several points in the transmission of the power from its generation to its application. A number of names are used for the power developed at various stages in this process, but none is a clear indicator of either the measurement system or definition used.

In the case of an engine dynamometre, power is measured at the engine's flywheel. With a chassis dynamometre or *rolling road*, power output is measured at the driving wheels. This accounts for the significant power loss through the drive train.

In General: Nominal is derived from the size of the engine and the piston speed and is only accurate at a pressure of 48 kPa (7 psi).

Indicated or gross horsepower (theoretical capability of the engine) [PLAN/ 33000] minus frictional losses within the engine (bearing drag, rod and crankshaft windage losses, oil film drag, etc.), equals

Brake / net / crankshaft horsepower (power delivered directly to and measured at the engine's crankshaft) minus frictional losses in the transmission (bearings, gears, oil drag, windage, etc.), equals

Shaft horsepower (power delivered to and measured at the output shaft of the transmission, when present in the system) minus frictional losses in the universal joint/s, differential, wheel bearings, tire and chain, (if present), equals Effective, True (thp) or commonly referred to as wheel horsepower (whp) All the above assumes that no power inflation factors have been applied to any of the readings.

Engine designers use expressions other than horsepower to denote objective targets or performance, such as brake mean effective pressure

(BMEP). This is a coefficient of theoretical brake horsepower and cylinder pressures during combustion.

Nominal Horsepower

Nominal horsepower (nhp) is an early 19th-century rule of thumb used to estimate the power of steam engines.

nhp = 7 x area of piston x equivalent piston speed/33,000

For paddle ships the piston speed was estimated as 129.7 x (stroke).

The stroke was the distance moved by the piston.

For the nominal horsepower to equal the actual power it would be necessary for the mean steam pressure in the cylinder during the stroke to be 48 kPa (7 psi) and for the piston speed to be of the order of 54–75 m/min.

The French Navy used the same definition of nominal horse power as Britain.

Indicated Gorsepower

Indicated horsepower (ihp) is the theoretical power of a reciprocating engine if it is completely frictionless in converting the expanding gas energy (piston pressure × displacement) in the cylinders. It is calculated from the pressures developed in the cylinders, measured by a device called an *engine indicator* – hence indicated horsepower. As the piston advances throughout its stroke, the pressure against the piston generally decreases, and the indicator device usually generates a graph of pressure vs stroke within the working cylinder. From this graph the amount of work performed during the piston stroke may be calculated.

Indicated horsepower was a better measure of engine power than nominal horsepower (nhp) because it took account of steam pressure. But unlike later measures such as shaft horsepower (shp) and brake horsepower (bhp), it did not take into account power losses due to transmission and gear boxes (if any).

Brake Horsepower

Brake horsepower (bhp) is the measure of an engine's horsepower before the loss in power caused by the gearbox and drive train. In Europe the DIN standard tested the engine fitted with all ancillaries and exhaust system as used in the car. The American SAE system tests without alternator, water pump, and other auxiliary components such as power steering pump, muffled exhaust system, etc. so the figures are higher than the European figures for the same engine. *Brake* refers to the device which was used to load an engine and hold it at a desired rotational speed. During testing, the output torque and rotational speed were measured to determine the *brake horsepower*. Horsepower was originally measured and calculated by use of the "indicator diagram" (a James Watt invention of the late 18th century), and later by means of a de Prony brake connected to the engine's output shaft.

More recently, an electrical brake dynamometre is used instead of a De Prony brake. Although the output delivered to the driving wheels is less than that obtainable at the engine's crankshaft, a chassis dynamometre gives an indication of an engine's "real world" horsepower after losses in the drive train and gearbox.

Shaft Horsepower

Shaft horsepower (shp) is the power delivered to the propeller shafts of a steamship (or one powered by diesel engines or nuclear power), or an aircraft powered by a piston engine or a gas turbine engine, and the rotors of a helicopter. This shaft horsepower can be measured with instruments, or estimated from the indicated horsepower and a standard figure for the losses in the transmission (typical figures are around 10%). This measure is not commonly used in the automobile industry, because in that context drive train losses can become significant.

Engine Power Test Codes

Engine power test codes determine how the power and torque of an automobile engine is measured and corrected. Correction factors are used to adjust power and torque measurements to standard atmospheric conditions to provide a more accurate comparison between engines as they are affected by the pressure, humidity, and temperature of ambient air. There exist several standards for this purpose, some described below.

Society of Automotive Engineers/SAE International

SAE Gross Power

Prior to the 1972 model year, American automakers rated and advertised their engines in brake horsepower (bhp), frequently referred to as SAE gross horsepower, because it was measured in accord with the protocols defined in SAE standards J245 and J1995. As with other brake horsepower test protocols, SAE gross hp was measured using a stock test engine, generally running with few belt-driven accessories and sometimes fitted with long tube test headers in lieu of the OEM exhaust manifolds. The atmospheric correction standards for barometric pressure, humidity and temperature for testing were relatively idealistic.

SAE Net Power

In the United States, the term *bhp* fell into disuse in 1971–72, as automakers began to quote power in terms of SAE net horsepower in accord with SAE standard J1349. Like SAE gross and other brake horsepower protocols, SAE Net hp is measured at the engine's crankshaft, and so does not account for transmission losses. However, the SAE net power testing protocol calls for standard production-type belt-driven accessories, air cleaner, emission controls, exhaust system, and other power-consuming accessories. This produces ratings in closer alignment with the power produced by the engine as it is actually configured and sold.

SAE Certified Power

In 2005, the SAE introduced "SAE Certified Power" with SAE J2723. This test is voluntary and is in itself not a separate engine test code but a certification of either J1349 or J1995 after which the manufacturer is allowed to advertise "Certified to SAE J1349" or "Certified to SAE J1995" depending on which test standard have been followed. To attain certification the test must follow the SAE standard in question, take place in an ISO9000/9002 certified facility and be witnessed by an SAE approved third party.

A few manufacturers such as Honda and Toyota switched to the new ratings immediately, with multi-directional results; the rated output of Cadillac's supercharged Northstar V8 jumped from 440 to 469 hp (328 to 350 kW) under the new tests, while the rating for Toyota's Camry 3.0 L *1MZ-FE* V6 fell from 210 to 190 hp (160 to 140 kW). The company's Lexus ES 330 and Camry SE V6 were previously rated at 225 hp (168 kW) but the ES 330 dropped to 218 hp (163 kW) while the Camry declined to 210 hp (160 kW). The first engine certified under the new program was the 7.0 L LS7 used in the 2006 Chevrolet Corvette Z06. Certified power rose slightly from 500 to 505 hp (373 to 377 kW).

While Toyota and Honda are retesting their entire vehicle lineups, other automakers generally are retesting only those with updated powertrains. For example, the 2006 Ford Five Hundred is rated at 203 horsepower, the same as that of 2005 model. However, the 2006 rating does not reflect the new SAE testing procedure as Ford is not going to spend the extra expense of retesting its existing engines. Over time, most automakers are expected to comply with the new guidelines.

SAE tightened its horsepower rules to eliminate the opportunity for engine manufacturers to manipulate factors affecting performance such as how much oil was in the crankcase, engine control system calibration, and whether an engine was tested with premium fuel. In some cases, such can add up to a change in horsepower ratings.

Deutsches Institut für Normung *70020*

DIN 70020 is a standard from German DIN regarding road vehicles. DIN testing, unlike SAE, tested the engine as installed in the vehicle, with cooling system, charging system and stock exhaust system all connected. Because the German word for *horsepower* is *Pferdestärke*, in Germany it is commonly abbreviated to *PS*. DIN hp is measured at the engine's output shaft, and is usually expressed in metric (Pferdestärke) rather than mechanical horsepower.

Economic Commission for Europe R24

ECE R24 is a UN standard for the approval of compression ignition engine emissions, installation and measurement of engine power. It is similar to DIN 70020 standard, but with different requirements for connecting an engine's fan during testing causing it to absorb less power from the engine.

Economic Commission for Europe R85

ECE R85 is a UN standard for the approval of internal combustion engines with regard to the measurement of the net power.

80/1269/EEC

80/1269/EEC of 16 December 1980 is a European Union standard for road vehicle engine power.

International Organisation for Standardization

- ISO 14396 specifies the additional and method requirement for determining the power of reciprocating internal combustion engines when presented for an ISO 8178 exhaust emission test. It applies to reciprocating internal combustion engines for land, rail and marine use excluding engines of motor vehicles primarily designed for road use.
- ISO 1585 is an engine net power test code intended for road vehicles.
- ISO 2534 is an engine gross power test code intended for road vehicles
- ISO 4164 is an engine net power test code intended for mopeds.

- ISO 4106 is an engine net power test code intended for motorcycles.
- ISO 9249 is an engine net power test code intended for earth moving machines.

Japanese Industrial Standard D 1001

JIS D 1001 is a Japanese net, and gross, engine power test code for automobiles or trucks having a spark ignition, diesel engine, or fuel injection engine.

What should be the Air-Fuel Ratio in Internal Combustion Engine

In internal combustion engine, atmospheric air is essentially required to burn the fuel. The Oxygen in air, helps for proper burning of fuel. To obtain excellent combustion result from fuel, the air and the fuel must be mixed in proper ratio. For complete combustion, the air-fuel ratio is approximately 15 : 1 by weight. This ratio is known as Chemically correct air-fuel ratio. This ideal ration in any internal combustion engine. However, the air-fuel ratio can be range from 20 : 1 to 8 : 1, in this range also combustion of fuel can occur. Any ratio outside of this range is either too rich or too lean to sustain flame propagation.

As per requirement of engine, the carburettor provide an air-fuel ratio, which must be within combustion range. Engine is cold at the time of starting so, very rich mixture is required. Rich mixture is also required at time of idling and producing maximum power. During the normal running, a comparatively lean mixture can be used.

For petrol engine, different air-fuel ratios are required under various conditions of load. These are as discussed below.

Air-Fuel Ratio for Starting

Very rich mixture (10 : 1) is required at starting of engine. During starting very small amount of fuel is vaporises and rest of it stay in the liquid state so as to give an ignitable mixture.

Air-Fuel Ratio for Idling

An idling, engine demands a rich mixture, which can be made leaner as the throttle is gradually opened. During idling, the pressure in the inlet manifold is about 20 to 25% of atmospheric pressure. At suction stroke, inlet valve opens and the product of combustion trapped in the clearance volume, expands in the inlet manifold. Latter when the piston moves downwards, the gases along with the fresh charges go into the cylinder. A rich mixture must be supplied during idling, to counteract the tendency of dilution and to get an ignitable mixture.

Air-Fuel Ratio for Medium Load

Most of the time, engine is running in medium load condition, therefore, it is desirable that the running should be most economical in this condition. So a lean mixture can be supplied, as engine has low fuel consumption at medium load. For multi cylinder engine, slightly more fuel is required due to mal distribution of fuel.

Air-Fuel Ratio for Maximum Power Range

When maximum power is required, the engine must be supplied with rich mixture as the economy is of no consideration. As the engine enters in the power range, the spark must be retarded otherwise knocking would occur.

A lean mixture burns at latter part of working stroke. As the exhaust valve expose to high temperature gases and have very less time to cool down. Moreover, the excess air in the lean mixture may cause an oxidising action on the hot exhaust valve and leads to failure.

Air-Fuel Ratio for Acceleration

Even during normal running, sometimes more power is required for a short period such as to accelerate the vehicle for overtaking etc. During this period rich mixture is required.

Different Types of Valves Used in Internal Combustion Engine.

In four stroke internal combustion engine, the "Poppet Valve" performed the opening of the cylinder to inlet or exhaust manifold at the correct moment. Generally the face of valve is ground at 45 degree but in same cases it is ground at 30 degree also.

It is not important to have a same angle of face in inlet and exhaust valve of same engines. To make it in right order, the valve may be reground after some use. There is some margin provided to avoid sharp edges.

The groove, retain the valve spring which aids in keeping the valve pressed against the seat when closed and thus seal the combustion space tightly. In close position, the valve face, fits the accurately matched ground seat in the cylinder block. Generally replaceable ring inserts are used for exhaust valve seat.

The inlet valves are made from plain nickel, nickel chrome or chrome molybdenum. Where as exhaust valves are made from nickel chrome, silicon chrome steel, high speed steel, stainless steel, high nickel chrome, tungsten steel and cobalt chrome steel.

Valve Mechanism of a Side Valve

Poppet valve has following main parts,

1. Cam Shaft
2. Cam
3. Cam Follower
4. Tappet
5. Adjusting Screw
6. Washer
7. Valve Spring
8. Valve Stem
9. Valve Stem Guide
10. Valve Face

With the help of these parts, valve performs it's operation very accurately in internal combustion engine.

The cam actuates the movement of the valve through the tappet. The replaceable valve stem moves up and down in the valve stem guide. This movement is obtained by rotation of camshaft and cam, which generally runs at the half the engine speed.

The valve spring, keeps the valve pressed against its seat and ensure a leakage proof operation and also bring back the valve very quickly during its closing.

When the engine is started, it gets heated up gradually there by causing the valve stem to expands. A valve tappet clearance is always provided to allow the expansion of valve stem and other parts. This clearance value depends upon the length of the valve, its material and the operating temperature of the engine. The tappet valve clearance can be adjusted by rotating the adjusting screw.

Where adjusting screw is not provided to vary the clearance, it can be increased by grinding the bottom of the valve stem and face or by using longer valve. Due care must be taken because even a slightly insufficient clearance may lead to the valve not properly resting against its seat as the engine gets heated causing increased noise level and loss of power.

The clearance provided in exhaust valve is slightly more than that of inlet valve. This is due to slightly more expansion in exhaust valve because of higher temperature of hot exhaust gases produced during combustion.

Valve Mechanism of Overhead Valve: In this mechanism, a push rod and a rocker arm, to push the valve against the spring pressure is required. The rocker arm rotates about the rocker arm shaft under the force exerted by the push rod. The clearance in this case is kept between the rocker arm and the valve stem and can be adjusted by the screw adjuster. A suitable gear train or chain is provided to convert the drive from crankshaft to camshaft.

In high speed engines, the frequency of vibration of the valve spring coincides with normal operating frequency of the valve which leads to resonance and increasing the surging effect. To overcome this problem, now a days compound spring are used. Compound spring means "One spring within the other of different natural frequencies".

Valves are the most important part of every engines. So, due care must be taken in selection and maintenance of valve.

How to Cool Internal Combustion Engine by Water?

To pass some of the heat generated due to combustion in the cylinder of internal combustion engine, air cooling or water cooling methods are used. The details about air cooling were covered in previous post (How To Cool Motor Cycle Engine?). In this post you will find some more methods of water cooling.

In water cooling method, the advantage of superior convective and conductive properties of water is used. Water is circulated continuously through the cylinder with an annular space known as Water Jacket. To avoid unequal expansion in the cylinder bore and burning of lubricating oil, the water jackets are so designed that they will cover the entire length of piston stroke.

For the cleaning of water jackets in large cylinders, cleaning doors are provided. This method is also employed in large reciprocating air compressor where we have to add cooling tower in addition to cool the circulating water.

Water Cooling can be done by any of the following methods,

1. *Direct Method:* In this method, water from a storage tank is directly supplied through an inlet valve to the engine cylinder. The hot water is simply discharged and not cooled for reuse. It is suitable for large industrial units and where plenty of water is easily available.
2. *Thermosyphon Method:* In this method, hot water from engine cylinder flows towards radiator which has comparatively cold water and get cool down. In order to ensure that the coolest

water is always made available to the water jackets the latter should be located at as low a level as possible with respect to the radiator.

This system is useful due its simplicity and automatic operation, but there is disadvantage of water freezing in cold weather condition. We have to use large quantity of water as there is slow rate of circulation observed.

Impeller Thermosyphon Method

Here the flow of water takes place by convection assisted by a pump. The latter is made non-positive type which even when stationary; allows the flow of water to take place. Thus when the working of pump fails water can still circulate on the thermosyphon system.

In its simplest form, water flow from radiator to engine and back to radiator. A thermostat is employed in system which prevents the flow of water below a certain temperature, from the engine to the radiator. This we have to do to acquire some of the temperature at starting of the engine.

This system has three major parts, (The details about this parts will be covered in coming post.)

1. Radiator
2. Fan and Pump
3. Thermostat
4. *Full Pump Circulation Method*

In this method, a positive supply of water by a centrifugal pump placed in the system. This system works on higher velocity of circulating water due to which less quantity of water is required hence we can use smaller radiator.

In some designs the piston is cooled by oil squirted against the piston crown underside through a nozzle located in the connecting rod small end.

Evaporative Cooling Method

In this method, the engine may be cooled by evaporating the water in the cylinder jackets, into the steam which absorbs large quantity of heat and raise to the top of the engine from where it flows into a tank at the bottom of the radiator and then flows upwards and gets condensed before reaching the top.

Due to large quantity of latent heat absorbed during evaporation of water into steam the weight of the circulating water is only 40% of that in other water cooling methods. Hence smaller radiator can be employed.

This system is quite useful where plenty of water is not available. The higher running temperature no doubt reduce the friction in the piston but the volumetric efficiency is reduced and the engine is also liable to detonate.

How You can Connect Piston to Crankshaft in Internal Combustion Engine?

In every internal combustion engine, there are various parts, connected to each other. To convert reciprocating motion of piston into rotary motion of crankshaft, we have to connect, piston to crankshaft with the help of connecting rod and gudgeon pin. Below you will find details about these two parts.

Connecting Rod

The load on the piston due to combustion of fuel in the combustion chamber is transmitted to crankshaft through the connecting rod. One end of connecting rod known as small end and is connected to the piston through gudgeon pin while the other end known as big end and is connected to crankshaft through crank pin.

Connecting rods are usually made up of drop forged I section. In large size internal combustion engine, the connecting rods of rectangular section have been employed. In such cases, the larger dimensions are kept in the plane of rotation.

In petrol engine, the connecting rod's big end is generally split to enable its clamping around the crankshaft. Suitable diameter holes are provided to accommodate connecting rod bolts for clamping. The big end of connecting rod is clamped with crankshaft with the help of connecting rod bolt, nut and split pin or cotter pin.

Generally, plain carbon steel is used as material to manufacture connecting rod but where low weight is most important factor, aluminium alloys are most suitable. Nickel alloy steel are also used for heavy duty engine's connecting rod.

Gudgeon Pin

This pin connects the piston with small end of the connecting rod and also known as piston pin. It is made up of case hardened steel and accurately ground to the required diameters. Gudgeon pins are made hollow to reduce its weight, resulting low inertia effect of reciprocating parts.

This pin is also known as "Fully Floating" as this is free to turn or oscillate both in the piston bosses as well as the small end of the connecting rod. There are very less chances of seizure in this case but the end movement of the pin must be restricted to score the cylinder walls. This can be achieved by using any one of the following three methods,

A) One spring circlip at each end is fitted into the groove in the piston bosses.

B) On spring circlip is provided in the middle.

C) Bronze or Aluminium pads are fitted at both ends of the pin, which prevents the cylinder walls from being damaged.

The gudgeon pin may also be semi-floating type, in which either the pin is free to turn or oscillate in the small end bearing but secured in the piston bosses or it may secured in the small end bearing and allowed a free oscillating movement in the piston bosses. This method provides more bearing area at the bosses and hence no need for providing bushes there in, is preferred.

Chapter 8

Fuel Injection

Fuel injection is a system for admitting fuel into an internal combustion engine. It has become the primary fuel delivery system used in automotive engines, having replaced carburetors during the 1980s and 1990s. A variety of injection systems have existed since the earliest usage of the internal combustion engine.

The primary difference between carburetors and fuel injection is that fuel injection atomizes the fuel by forcibly pumping it through a small nozzle under high pressure, while a carburetor relies on suction created by intake air accelerated through a Venturi tube to draw the fuel into the airstream.

Modern fuel injection systems are designed specifically for the type of fuel being used. Some systems are designed for multiple grades of fuel (using sensors to adapt the tuning for the fuel currently used). Most fuel injection systems are for gasoline or diesel applications.

Objectives

The functional objectives for fuel injection systems can vary. All share the central task of supplying fuel to the combustion process, but it is a design decision how a particular system is optimized. There are several competing objectives such as:

- Power output
- Fuel efficiency
- Emissions performance
- Ability to accommodate alternative fuels
- Reliability

- Driveability and smooth operation
- Initial cost
- Maintenance cost
- Diagnostic capability
- Range of environmental operation
- Engine tuning

The modern digital electronic fuel injection system is more capable at optimizing these competing objectives consistently than earlier fuel delivery systems (such as carburetors). Carburetors have the potential to atomize fuel better.

Benefits

Driver Benefits

Operational benefits to the driver of a fuel-injected car include smoother and more dependable engine response during quick throttle transitions, easier and more dependable engine starting, better operation at extremely high or low ambient temperatures, smoother engine idle and running, increased maintenance intervals, and increased fuel efficiency. On a more basic level, fuel injection does away with the choke, which on carburetor-equipped vehicles must be operated when starting the engine from cold and then adjusted as the engine warms up.

Environmental Benefits

Fuel injection generally increases engine fuel efficiency. With the improved cylinder-to-cylinder fuel distribution of multi-point fuel injection, less fuel is needed for the same power output (when cylinder-to-cylinder distribution varies significantly, some cylinders receive excess fuel as a side effect of ensuring that all cylinders receive *sufficient* fuel).

Exhaust emissions are cleaner because the more precise and accurate fuel metring reduces the concentration of toxic combustion byproducts leaving the engine, and because exhaust cleanup devices such as the catalytic converter can be optimized to operate more efficiently since the exhaust is of consistent and predictable composition.

History and Development

Herbert Akroyd Stuart developed the first device with a design similar to modern fuel injection, using a 'jerk pump' to metre out fuel oil at high pressure to an injector. This system was used on the hot bulb engine and was adapted and improved by Bosch and Clessie Cummins for use on diesel engines (Rudolf Diesel's original system employed a cumbersome 'air-blast' system using highly compressed air). Fuel injection was in widespread commercial use in diesel engines by the mid-1920s.

An early use of indirect gasoline injection dates back to 1902, when French aviation engineer Leon Levavasseur pioneered it on his Antoinette 8V aircraft powerplant.

Another early use of gasoline direct injection (i.e. injection of gasoline, also known as petrol) was on the Hesselman engine invented by Swedish engineer Jonas Hesselman in 1925. Hesselman engines use the ultra lean burn principle; fuel is injected toward the end of the compression stroke, then ignited with a spark plug. They are often started on gasoline and then switched to diesel or kerosene.

Direct fuel injection was used in notable World War II aero-engines such as the Junkers Jumo 210, the Daimler-Benz DB 601, the BMW 801, the Shvetsov ASh-82FN (M-82FN). German direct injection petrol engines used injection systems developed by Bosch from their diesel injection systems. Later versions of the Rolls-Royce Merlin and Wright R-3350 used single point fuel injection, at the time called "Pressure Carburettor". Due to the wartime relationship between Germany and Japan, Mitsubishi also had two radial aircraft engines utilising fuel injection, the Mitsubishi Kinsei (*kinsei* means "venus") and the Mitsubishi Kasei (*kasei* means "mars").

Alfa Romeo tested one of the very first electronic injection systems (Caproni-Fuscaldo) in Alfa Romeo 6C2500 with "Ala spessa" body in 1940 Mille Miglia. The engine had six electrically operated injectors and were fed by a semi-high-pressure circulating fuel pump system.

Development in Diesel Engines

All diesel engines (with the exception of some tractors and scale model engines have fuel injected into the combustion chamber.

Development in Gasoline/Petrol Engines

Mechanical Injection

The invention of mechanical injection for gasoline-fuelled aviation engines was by the French inventor of the V8 engine configuration, Leon Levavasseur in 1902. Levavasseur designed the original Antoinette firm's series of V-form aero engines, starting with the Antoinette 8V to be used by the aircraft the Antoinette firm built that Levavasseur also designed, flown from 1906 to the firm's demise in 1910, with the world's first V16 engine, using Levavasseur's direct injection and producing some 100 hp, flying an Antoinette VII monoplane in 1907.

The first post-World War I example of direct gasoline injection was on the Hesselman engine invented by Swedish engineer Jonas Hesselman in 1925. Hesselman engines used the ultra lean burn principle and injected the fuel in the end of the compression stroke and then ignited it with a spark plug, it was often started on gasoline and then switched over to run on diesel or kerosene. The Hesselman engine was a low compression design constructed to run on heavy fuel oils.

Direct gasoline injection was applied during the Second World War to almost all higher-output production aircraft powerplants made in Germany (the widely used BMW 801 radial, and the popular inverted inline V12 Daimler-Benz DB 601, DB 603 and DB 605, along with the similar Junkers Jumo 210G, Jumo 211 and Jumo 213, starting as early as 1937 for both the Jumo 210G and DB 601), the Soviet Union (Shvetsov ASh-82FN radial, 1943, Chemical Automatics Design Bureau - KB Khimavtomatika) and the USA (Wright R-3350 *Duplex Cyclone* radial, 1944).

Immediately following the war, hot rodder Stuart Hilborn started to offer mechanical injection for race cars, salt cars, and midgets, well-known and easily distinguishable because of their prominent velocity stacks projecting upwards from the engine they were used on.

The first automotive direct injection system used to run on gasoline was developed by Bosch, and was introduced by Goliath for their Goliath GP700 automobile, and Gutbrod in 1952. This was basically a high-pressure diesel direct-injection pump with an intake throttle valve set up. (Diesels only change the amount of fuel injected to vary output; there is no throttle.) This system used a normal gasoline fuel pump, to provide fuel to a mechanically driven injection pump, which had separate plungers per injector to deliver a very high injection pressure directly into the combustion chamber. The 1954 Mercedes-Benz W196 Formula 1 racing car engine used Bosch direct injection derived from wartime aero engines. Following this racetrack success, the 1955 Mercedes-Benz 300SL, the first production sports car to use fuel injection, used direct injection. The same engine was used in the Mercedes-Benz 300SLR famously driven by Stirling Moss to victory in the 1955 Mille Miglia. The Bosch fuel injectors were placed into the bores on the cylinder wall used by the spark plugs in other Mercedes-Benz six-cylinder engines (the spark plugs were relocated to the cylinder head). Later, more mainstream applications of fuel injection favoured the less-expensive indirect injection methods.

Chevrolet introduced a mechanical fuel injection option, made by General Motors' Rochester Products division, for its 283 V8 engine in 1956 (1957 US model year). This system directed the inducted engine air across a "spoon shaped" plunger that moved in proportion to the air volume. The plunger connected to the fuel metring system that mechanically dispensed fuel to the cylinders via distribution tubes. This system was not a "pulse" or intermittent injection, but rather a constant flow system, metring fuel to all cylinders simultaneously from a central "spider" of injection lines. The fuel metre adjusted the amount of flow according to engine speed and load, and included a fuel reservoir, which was similar to a carburetor's float chamber. With its own high-pressure fuel pump driven by a cable from the distributor to the fuel metre, the system supplied the necessary pressure for injection. This was a "port" injection where the injectors are located in the intake manifold, very near the intake valve.

During the 1960s, other mechanical injection systems such as Hilborn were occasionally used on modified American V8 engines in various racing applications such as drag racing, oval racing, and road racing. These racing-derived systems were not suitable for everyday street use, having no provisions for low speed metring, or often none even for starting (starting required that fuel be squirted into the injector tubes while cranking the engine). However, they were a favourite in the aforementioned competition trials in which essentially wide-open throttle operation was prevalent. Constant-flow injection systems continue to be used at the highest levels of drag racing, where full-throttle, high-RPM performance is key.

Another mechanical system, made by Bosch called Jetronic, but injecting the fuel into the port above the intake valve, was used by several European car makers, particularly Porsche from 1969 until 1973 in the 911 production range and until 1975 on the Carrera 3.0 in Europe. Porsche continued using this system on its racing cars into the late seventies and early eighties. Porsche racing variants such as the 911 RSR 2.7 & 3.0, 904/6, 906, 907, 908, 910, 917 (in its regular normally aspirated or 5.5 Liter/1500 HP Turbocharged form), and 935 all used Bosch or Kugelfischer built variants of injection. The early Bosch Jetronic systems were also used by Audi, Volvo, BMW, Volkswagen, and many others. The Kugelfischer system was also used by the BMW 2000/2002 Tii and some versions of the Peugeot 404/504 and Lancia Flavia. Lucas also offered a mechanical system that was used by some Maserati, Aston Martin, and Triumph models between 1963 and 1973.

A system similar to the Bosch inline mechanical pump was built by SPICA for Alfa Romeo, used on the Alfa Romeo Montreal and on U.S. market 1750 and 2000 models from 1969 to 1981. This was designed to meet the U.S. emission requirements with no loss in performance and it also reduced fuel consumption.

Electronic Injection

The first commercial electronic fuel injection (EFI) system was Electrojector, developed by the Bendix Corporation and was offered by American Motors Corporation (AMC) in 1957. The Rambler Rebel, showcased AMC's new 327 cu in (5.4 L) engine. The Electrojector was an option and rated at 288 bhp (214.8 kW). The EFI produced peak torque 500 rpm lower than the equivalent carburetored engine The Rebel Owners Manual described the design and operation of the new system. (due to cooler, therefore denser, intake air).

The cost of the EFI option was US$395 and it was available on 15 June 1957. Electrojector's teething problems meant only pre-production cars were so equipped: thus, very few cars so equipped were ever sold and none were made available to the public. The EFI system in the Rambler ran fine in warm weather, but suffered hard starting in cooler temperatures.

Chrysler offered Electrojector on the 1958 Chrysler 300D, DeSoto Adventurer, Dodge D-500 and Plymouth Fury, arguably the first series-production cars equipped with an EFI system. It was jointly engineered by Chrysler and Bendix. The early electronic components were not equal to the rigors of underhood service, however, and were too slow to keep up with the demands of "on the fly" engine control. Most of the 35 vehicles originally so equipped were field-retrofitted with 4-barrel carburetors. The Electrojector patents were subsequently sold to Bosch.

Bosch developed an electronic fuel injection system, called *D-Jetronic* (*D* for *Druck,* German for "pressure"), which was first used on the VW 1600TL/E in 1967. This was a speed/density system, using engine speed and intake manifold air density to calculate "air mass" flow rate and thus fuel requirements. This system was adopted by VW, Mercedes-Benz, Porsche, Citroën, Saab, and Volvo. Lucas licensed the system for production with Jaguar.

Bosch superseded the D-Jetronic system with the *K-Jetronic* and *L-Jetronic* systems for 1974, though some cars (such as the Volvo 164) continued using D-Jetronic for the following several years. In 1970, the Isuzu 117 Coupé was introduced with a Bosch-supplied D-Jetronic fuel injected engine sold only in Japan.

Figure: *Chevrolet Cosworth Vega engine showing Bendix electronic fuel injection (in orange).*

In Japan, the Toyota Celica used electronic, multi-port fuel injection in the optional 18R-E engine in January 1974. Nissan offered electronic, multi-port fuel injection in 1975 with the Bosch L-Jetronic system used in the Nissan L28E engine and installed in the Nissan Fairlady Z, Nissan Cedric, and the Nissan Gloria. Nissan also installed multi-point fuel injection in the Nissan Y44 V8 engine in the Nissan President. Toyota soon followed with the same technology in 1978 on the 4M-E engine installed in the Toyota Crown, the Toyota Supra, and the Toyota Mark II. In the 1980s, the Isuzu Piazza, and the Mitsubishi Starion added fuel injection as standard equipment, developed separately with both companies history of diesel powered engines. 1981 saw Mazda offer fuel injection in the Mazda Luce with the Mazda FE engine, and in 1983, Subaru offered fuel injection in the Subaru EA81 engine installed in the Subaru Leone. Honda followed in 1984 with their own system, called PGM-FI in the Honda Accord, and the Honda Vigor using the Honda ES3 engine.

The limited production Chevrolet Cosworth Vega was introduced in March 1975 using a Bendix EFI system with pulse-time manifold injection, four injector valves, an electronic control unit (ECU), five independent sensors and two fuel pumps. The EFI system was developed to satisfy stringent emission control requirements and market demands for a technologically advanced responsive vehicle. 5000 hand-built Cosworth Vega engines were produced but only 3,508 cars were sold through 1976.

The Cadillac Seville was introduced in 1975 with an EFI system made by Bendix and modelled very closely on Bosch's D-Jetronic. L-Jetronic first appeared on the 1974 Porsche 914, and uses a mechanical airflow metre (L for *Luft*, German for "air") that produces a signal that is proportional to "air volume". This approach required additional sensors to measure the atmospheric pressure and temperature, to ultimately calculate "air mass". L-Jetronic was widely adopted on European cars of that period, and a few Japanese models a short time later.

In 1980, Motorola (now Freescale) introduced the first electronic engine control unit, the EEC-III. Its integrated control of engine functions (such as fuel injection and spark timing) is now the standard approach for fuel injection systems. The Motorola technology was installed in Ford North American products.

Supersession of Carburetors

In the 1970s and 1980s in the US and Japan, the respective federal governments imposed increasingly strict exhaust emission regulations. During that time period, the vast majority of gasoline-fuelled automobile and light truck engines did not use fuel injection. To comply with the new regulations, automobile manufacturers often made extensive and complex modifications to the engine carburetor(s). While a simple carburetor system is cheaper to manufacture than a fuel injection system, the more complex carburetor systems installed on many engines in the 1970s were much more costly than the earlier simple carburetors. To more easily comply with emissions regulations, automobile manufacturers began installing fuel injection systems in more gasoline engines during the late 1970s.

The open loop fuel injection systems had already improved cylinder-to-cylinder fuel distribution and engine operation over a wide temperature range, but did not offer further scope to sufficient control fuel/air mixtures, in order to further reduce exhaust emissions. Later Closed loop fuel injection systems improved the air/fuel mixture control with an exhaust gas oxygen sensor and began incorporating a catalytic converter to further reduce exhaust emissions. Fuel injection was phased in through the latter 1970s and 80s at an accelerating rate, with the German, French, and U.S. markets leading and the UK and Commonwealth markets lagging somewhat. Since the early 1990s, almost all gasoline passenger cars sold in first world markets are equipped with electronic fuel injection (EFI). The carburetor remains in use in developing countries where vehicle emissions are unregulated and diagnostic and repair infrastructure is sparse. Fuel injection is gradually replacing carburetors in these nations too as they adopt emission regulations conceptually similar to those in force in Europe, Japan, Australia, and North America.

Many motorcycles still utilise carburetored engines, though all current high-performance designs have switched to EFI.

NASCAR finally replaced carburetors with fuel-injection, starting at the beginning of the 2012 NASCAR Sprint Cup Series season.

System Components

System Overview

The process of determining the necessary amount of fuel, and its delivery into the engine, are known as fuel metring. Early injection systems used mechanical methods to metre fuel, while nearly all modern systems use electronic metring.

Determining How Much Fuel to Supply

The primary factor used in determining the amount of fuel required by the engine is the amount (by weight) of air that is being taken in by the engine for use in combustion. Modern systems use a mass airflow sensor to send this information to the engine control unit.

Data representing the amount of power output desired by the driver (sometimes known as "engine load") is also used by the engine control unit in calculating the amount of fuel required. A throttle position sensor (TPS) provides this information. Other engine sensors used in EFI systems include a coolant temperature sensor, a camshaft or crankshaft position sensor (some systems get the position information from the distributor), and an oxygen sensor which is installed in the exhaust system so that it can be used to determine how well the fuel has been combusted, therefore allowing closed loop operation.

Supplying the Fuel to the Engine

Fuel is transported from the fuel tank (via fuel lines) and pressurised using fuel pump(s). Maintaining the correct fuel pressure is done by a fuel pressure regulator. Often a fuel rail is used to divide the fuel supply into the required number of cylinders. The fuel injector injects liquid fuel into the intake air (the location of the fuel injector varies between systems).

EFI Gasoline Engine Components

Note: These examples specifically apply to a modern EFI gasoline engine. Parallels to fuels other than gasoline can be made, but only conceptually.

- Injectors
- Fuel Pump
- Fuel Pressure Regulator
- Engine control unit
- Wiring Harness

- Various Sensors (Some of the sensors required are listed here.)
- Crank/Cam Position: Hall effect sensor
- Airflow: MAF sensor, sometimes this is inferred with a MAP sensor
- Exhaust Gas Oxygen: oxygen sensor, EGO sensor, UEGO sensor

Engine Control Unit

The engine control unit is central to an EFI system. The ECU interprets data from input sensors to, among other tasks, calculate the appropriate amount of fuel to inject.

Fuel Injector

When signalled by the engine control unit the fuel injector opens and sprays the pressurised fuel into the engine. The duration that the injector is open (called the pulse width) is proportional to the amount of fuel delivered. Depending on the system design, the timing of when injector opens is either relative each individual cylinder (for a sequential fuel injection system), or injectors for multiple cylinders may be signalled to open at the same time (in a batch fire system).

Target Air/Fuel Ratios

The relative proportions of air and fuel vary according to the type of fuel used and the performance requirements (i.e. power, fuel economy, or exhaust emissions).

Various Injection Schemes

Single-Point Injection

Single-point injection uses a single injector at the throttle body (the same location as was used by carburetors).

It was introduced in the 1940s in large aircraft engines (then called the pressure carburetor) and in the 1980s in the automotive world (called Throttle-body Injection by General Motors, Central Fuel Injection by Ford, PGM-CARB by Honda, and EGI by Mazda). Since the fuel passes through the intake runners (like a carburetor system), it is called a "wet manifold system".

The justification for single-point injection was low cost. Many of the carburetor's supporting components- such as the air cleaner, intake manifold, and fuel line routing- could be reused. This postponed the redesign and tooling costs of these components. Single-point injection was used extensively on American-made passenger cars and light trucks during 1980-1995, and in some European cars in the early and mid-1990s.

Continuous Injection

In a continuous injection system, fuel flows at all times from the fuel injectors, but at a variable flow rate. This is in contrast to most fuel injection systems, which provide fuel during short pulses of varying duration, with a constant rate of flow during each pulse. Continuous injection systems can be multi-point or single-point, but not direct.

The most common automotive continuous injection system is Bosch's K-Jetronic, introduced in 1974. K-Jetronic was used for many years between 1974 and the mid-1990s by BMW, Lamborghini, Ferrari, Mercedes-Benz, Volkswagen, Ford, Porsche, Audi, Saab, DeLorean, and Volvo. Chrysler used a continuous fuel injection system on the 1981-1983 Imperial.

In piston aircraft engines, continuous-flow fuel injection is the most common type. In contrast to automotive fuel injection systems, aircraft continuous flow fuel injection is all mechanical, requiring no electricity to operate. Two common types exist: the Bendix RSA system, and the TCM system. The Bendix system is a direct descendant of the pressure carburetor. However, instead of having a discharge valve in the barrel, it uses a *flow divider* mounted on top of the engine, which controls the discharge rate and evenly distributes the fuel to stainless steel injection lines to the intake ports of each cylinder.

The TCM system is even more simple. It has no venturi, no pressure chambers, no diaphragms, and no discharge valve. The control unit is fed by a constant-pressure fuel pump. The control unit simply uses a butterfly valve for the air, which is linked by a mechanical linkage to a rotary valve for the fuel. Inside the control unit is another restriction, which controls the fuel mixture. The pressure drop across the restrictions in the control unit controls the amount of fuel flow, so that fuel flow is directly proportional to the pressure at the flow divider. In fact, most aircraft that use the TCM fuel injection system feature a fuel flow gauge that is actually a pressure gauge calibrated in *gallons per hour* or *pounds per hour* of fuel.

Central Port Injection

From 1992 to 1996 General Motors implemented a system called Central Port Injection or Central Port Fuel Injection. The system uses tubes with poppet valves from a central injector to spray fuel at each intake port rather than the central throttle-body. Fuel pressure is similar to a single-point injection system. CPFI (used from 1992 to 1995) is a batch-fire system, while CSFI (from 1996) is a sequential system.

Multiport Fuel Injection

Multiport fuel injection injects fuel into the intake ports just upstream of each cylinder's intake valve, rather than at a central point within an intake manifold. MPFI (or just MPI) systems can be sequential, in which injection is timed to coincide with each cylinder's intake stroke; batched, in which fuel is injected to the cylinders in groups, without precise synchronization to any particular cylinder's intake stroke; or simultaneous, in which fuel is injected at the same time to all the cylinders. The intake is only slightly wet, and typical fuel pressure runs between 40-60 psi.

Many modern EFI systems utilise sequential MPFI; however, in newer gasoline engines, direct injection systems are beginning to replace sequential ones.

Direct Injection

In a direct injection engine, fuel is injected into the combustion chamber as opposed to injection before the intake valve (petrol engine) or a separate pre-combustion chamber (diesel engine).

In a common rail system, the fuel from the fuel tank is supplied to the common header (called the accumulator). This fuel is then sent through tubing to the injectors, which inject it into the combustion chamber. The header has a high pressure relief valve to maintain the pressure in the header and return the excess fuel to the fuel tank. The fuel is sprayed with the help of a nozzle that is opened and closed with a needle valve, operated with a solenoid. When the solenoid is not activated, the spring forces the needle valve into the nozzle passage and prevents the injection of fuel into the cylinder. The solenoid lifts the needle valve from the valve seat, and fuel under pressure is sent in the engine cylinder. Third-generation common rail diesels use piezoelectric injectors for increased precision, with fuel pressures up to 1,800 bar or 26,000 psi.

Direct fuel injection costs more than indirect injection systems: the injectors are exposed to more heat and pressure, so more costly materials and higher-precision electronic management systems are required.

Diesel Engines

Most diesel engines (with the exception of some tractors and scale model engines) have fuel injected into the combustion chamber.

Earlier systems, relying on simpler injectors, often injected into a sub-chamber shaped to swirl the compressed air and improve combustion; this was known as indirect injection. However, this was less efficient than the now common direct injection in which initiation of combustion takes place in a depression (often toroidal) in the crown of the piston.

Throughout the early history of diesels, they were always fed by a mechanical pump with a small separate chamber for each cylinder, feeding separate fuel lines and individual injectors. Most such pumps were in-line, though some were rotary.

Most modern diesel engines use common rail or unit injector direct injection systems.

Gasoline Engines

Modern gasoline engines also utilise direct injection, which is referred to as gasoline direct injection. This is the next step in evolution from multi-point fuel injection, and offers another magnitude of emission control by eliminating the "wet" portion of the induction system along the inlet tract.

By virtue of better dispersion and homogeneity of the directly injected fuel, the cylinder and piston are cooled, thereby permitting higher compression ratios and earlier ignition timing, with resultant enhanced power output. More precise management of the fuel injection event also enables better control of emissions. Finally, the homogeneity of the fuel mixture allows for leaner air/fuel ratios, which together with more precise ignition timing can improve fuel efficiency. Along with this, the engine can operate with stratified (lean burn) mixtures, and hence avoid throttling losses at low and part engine load. Some direct-injection systems incorporate piezoelectronic fuel injectors. With their extremely fast response time, multiple injection events can occur during each cycle of each cylinder of the engine.

Swirl Injection

Swirl injectors are used in liquid rocket, gas turbine, and diesel engines to improve atomization and mixing efficiency.

The circumferential velocity component is first generated as the propellant enters through helical or tangential inlets producing a thin, swirling liquid sheet. A gas-filled hollow core is then formed along the centreline inside the injector due to centrifugal force of the liquid sheet. Because of the presence of the gas core, the discharge coefficient is generally low. In swirl injector, the spray cone angle is controlled by the ratio of the circumferential velocity to the axial velocity and is generally wide compared with nonswirl injectors.

Maintenance Hazards

Fuel injection introduces potential hazards in engine maintenance due to the high fuel pressures used. Residual pressure can remain in the fuel lines long after an injection-equipped engine has been shut down. This residual pressure must be relieved, and if it is done so by external bleed-off, the fuel must be safely contained. If a high-pressure diesel fuel injector is removed from its seat and operated in open air, there is a risk to the operator of injury by hypodermic jet-injection, even with only 100 psi (6.9 bar) pressure. The first known such injury occurred in 1937 during a diesel engine maintenance operation.

Engine Auxiliary Systems

Some equipment can be built on the engine, and the rest can be delivered separately or grouped in modules.

Depending on the engine type and application, lubricating oil pump, HT- and LT-cooling water pumps, fuel pump, oil filters and coolers, pre-lubricating oil pump and thermostatic valves can be built on the engine.

Stand-by pumps, seawater pumps, central coolers, starting air vessels, lubricating oil automatic filters, exhaust gas silencers and boilers are typically delivered for separate mounting.

Standardised Modular Auxiliary Units

- Fuel oil booster
- Fuel oil separating
- Lubricating oil separating
- Cooling water preheating
- Starting air compressors
- Oil mist separator
- Oily water bilge separator

Maximum compatibility is ensured when auxiliary systems are delivered together with the main propulsion engines and diesel generator sets. Whenever necessary, the auxiliary systems are tailored to optimise the operating performance for a specific trade. The systems are specified to minimise building costs and operating costs for a specific combination of main and auxiliary engines.

Farm Implement Maintenance and Repair

Planting season isn't just one of the toughest times of the year for the farmer, it's also one of the toughest times of the year for equipment.

And there's not much that's as frustrating as time lost to a broken implement.

Whether it's a bent haybine jack, a crack in a combine hopper or a dozer frame that snapped in two, equipment breakages can be more than a simple nuisance, they can keep you from harvesting your crops and even endanger your safety.

In an ideal world, your equipment would break at the end of the day right next to the shop so you could just take it back into the shop and fix it properly and quickly and get it ready for the next morning.

In the real world, however, metal brackets and jacks are more likely to follow Murphy's Law than your planting schedule and break down at the most inopportune moment and as far as possible from the shop. In those situations, Chris Roehl, a product manager with Miller Electric Mfg. Co. offers a racing analogy as a general rule of thumb for deciding how to approach the repair.

> *"In the middle of a race, a pit row mechanic is going to do whatever it takes to get that car back on the track," Roehl explains. "So if you can get a welding generator out there, make a quick weld and keep harvesting, that's what you should do. Then, at the end of the day, you should bring the implement back to your shop and take some time making a proper repair. If there is a safety issue involved, though, you don't want to risk losing a limb or even your life for a couple more hours in the field."*

Joel Ort, a technician with Miller, agrees. He notes that fixing the crack or hole right the first time will save you from having to fix it again next harvest season.

> *"A lot of times I'll see equipment that developed a crack and people will think there is something wrong with their welding skills or their welder when the crack reappears," Ort said. "Usually, though, that happens because they failed to properly prepare the piece before welding."*

In order to avoid having to deal with recurring cracks, Roehl and Ort offer these suggestions for getting the repair right the first time.

Clean the Surface

"One of the most common problems farmers have in making good, solid welds is that they don't grind the dirt, oil, paint and corrosion from the metal before welding," Ort explains. "Stick welding is more forgiving of dirt than MIG or TIG welding, but you still need to have the metal relatively free of foreign material."

A common hand-held die grinder, wheel grinder or wire brush should suffice for most repairs. It is also important to clean off the area around the crack or hole so that contaminants don't make their way into the weld puddle or heat-affected zone and thereby weaken the weld.

Cleaning the material to be welded is extra important when welding aluminum, which might look clean but still have a layer of aluminum oxide on the surface. "Aluminum looks nice and shiny because it doesn't rust, but it forms a layer of aluminum oxide, which will make it very difficult to weld and trap impurities in the welds," Roehl said.

Establish a Good Ground

Welding equipment relies on electrical current making a complete circuit from the power source to the material and back to the power source. The stinger and electrode (in Stick welding) or just the wire (in MIG welding) provide one connection from the power source to the material; the ground clamp provides the other connection.

In addition to cleaning the material being welded, it's important to remove any paint and corrosion from the piece where the ground clamp will attach. "A lot of guys will have problems striking an arc and keeping it lit when it's simply a result of not cleaning the material enough to establish a good ground," Ort said. A better grounding connection is also achieved by placing the clamp as close to the weld location as possible.

Grind out Cracks and Holes

This is perhaps the most important and frequently overlooked step in making a lasting weld. "Typically, if something breaks or forms a crack, a lot of people will just take their MIG or Stick welder and just pull the trigger and weld over the crack," Roehl said. "Well, unfortunately, that crack may be a half-inch deep. When you just put a little weld over the surface, it might only penetrate 1/8-in. deep, so you're not penetrating all the way to the root."

To ensure adequate penetration and a lasting weld, Ort recommends grinding or drilling the crack out all the way to the other side of the metal, as well as just beyond the length of the crack on each side. "There are people who will say that completely drilling it out isn't necessary," Ort said, "but I would recommend it. Even where it's not cracked, there can be stress within the metal that could cause it to crack somewhere down the road."

Roehl noted that a similar procedure is advisable for holes worn or cracked into buckets and other implements that use thinner metal. "Let's say you just take a welder and fill in a hole the size of a half-dollar, but you don't do anything about the little spider cracks that form around it," Roehl said. "What happens is that those cracks continue to grow and cause other holes and problems."

Instead, Roehl recommends grinding out the metal around the hole to determine how fatigued it is. "If you're grinding and chips are flying all over the place, that means the metal was very brittle. To repair a half-inch hole you might need to weld a three-inch coupon over it.

Use Multiple Passes

Another source of recurring frustration for farmers is that, hoping to save time, they will weld slower thinking they are achieving greater penetration into the base metal. Not so, Roehl says.

"Don't hesitate to make multiple passes," Roehl advises. "What can happen when you have a deep crack to fill in is you go real slow so you put more filler metal in, but you're actually going to be welding on top of the puddle instead of getting into the base metal. The filler metal is just sitting on top of the base metal then; it's not penetrating it."

Stick welding produces a layer of material called "slag" that must be chipped and brushed off the weld after each pass. One of the biggest benefits to MIG welding is that it doesn't create slag that must be chipped off after each pass and the farmer can make multiple passes without any additional preparation.

Self-shielded wire, which can be fed through a MIG welder but has the advantage of being impervious to wind, also produces slag that must be chipped off before attempting multiple passes.

An important step to remember when welding aluminum is to fill in the concave crater at the end of the weld. If the crater is left too concave, it will start cracking up the length of the weld. To avoid crater cracking, Roehl recommends letting the puddle cool for two seconds when you get to the end of the weld and pulling the trigger again briefly to fill in the crater until it is flush with the rest of the weld or even sits a little higher.

Use the Right Welding Equipment

"Quarter-inch steel is like sheet metal on farms," Roehl notes, "so they're really not going to be adequately served by a 135 or 175 amp machine. A farmer would be able to use a smaller machine like that for multiple passes, but what will happen is they'll exceed its duty cycle and have to wait several minutes for it to cool down before they can make another pass."

For MIG welding, Roehl recommends a machine with a 200 or 250-amp output with a duty cycle of at least 40 percent at the rated output, such as the Millermatic 210 or 251.

***Figure:** This farmer is using a Bobcat™ 250 welding generator to power a Millermatic® 175 MIG welder equipped with flux cored wire for this hitch repair.*

If you want to be able to weld out in the field, a multi-process welding generator, such as the Bobcat™ 250, is able to Stick, MIG, flux cored or TIG weld (with the correct accessories) while at the same time providing 10,500 watts of power for lights or virtually any other tools you might need in the field.

Take Care of Your Welding Equipment

It's tough to keep your machinery in good shape if your welder isn't. Because welders vary by make, manufacturer and the type of welding they do, the best way to ensure your welder is operating at its peak is to follow the manufacturer's recommended maintenance schedule.

There are, however, some general tips that will help keep all welders and accessories in good condition.

For engine-powered welding generators, Ort said the oil, oil filter and air filter should be changed at least every 100 hours of operation. Some models require oil changes every 50 hours.

Farms that don't use their generators that much in a single year should change the oil, oil filter and air filter each spring. Spark plugs should be checked roughly every 200 hours of use and cables and hoses should be checked and replaced if broken every 500 hours. Farms who don't use their generators very much should also add a conditioner and stabilizer to the gas tank, or in the case of diesel engines, an anti-gelling agent in the winter.

For most welders, both engine powered and not, it's a good idea to take the cover off a couple times a year and blow off the dust and dirt that can build up inside. Due to the sophisticated components in most modern welders, especially MIG and TIG welders, specific problems with a welder should be sent to a certified technician for repair.

Safety

Welding can be a great solution to many common problems on the farm, but appropriate safety precautions — many of which are common sense — need to be taken to avoid injury.

Working with electricity is always a potential safety hazard, so before you weld, make sure all of your outlets, wires and electrical connections are in good working order and away from any water sources.

If you're making field repairs or your shop has hay or other combustible material on the floor, it's a good idea to wet it down or lay down a special welding blanket before welding.

While not exactly life threatening, burns from sparks, spatter or chipped slag are still far from fun, and arc flash from insufficient eye protection feels like sand in the eyes or can even blind you temporarily. Appropriate safety apparel should be worn at all times.

Know When not to Weld

Last but not least, there are also times when it just isn't practical to repair the part with a welder.

"Some hard facing repairs, such as a skid plate on a haybine, will cost more for the box of electrodes than it would cost to buy a new skid plate," explains Ort. "Hardfacing it would definitely help the part last longer, but it can get expensive, even when you do it yourself."

While no one wants to need a welder, particularly during the long days of the planting season, the reality is that metal does wear out and crack and being able to repair your own equipment can save you time and money compared to hiring a weld shop to perform the repair. Although you're likely to find yourself back in the shop with new cracks and holes next season, following these few simple guidelines should ensure that it won't be because of the same cracks and holes you tried fixing this year.

Vehicle and Equipment Maintenance and Repair

Day-to-day maintenance and repair activities keep farm machinery and vehicles safe and reliable. Maintenance activities include oil and filter changes, battery replacement, and repairs including light metal machining.

Potential wastes generated as a result of farm machinery and vehicle maintenance and repair activities can include used oil, spent fluids, spent batteries, asbestos brake pads and linings, metal machining wastes, spent organic solvents, and tires. These wastes have the potential to be released to the environment if not handled properly, stored in secure areas with secondary containment, and/or protected from exposure to weather. If released to the environment, the impact of these releases can be contamination of surface waters, groundwater, and soils, as well as toxic releases to the air.

Used Oil

The impact of oil changes can be minimized by preventing releases of used oil to the environment, and recycling or reusing used oil whenever possible. Spills can be prevented by using containment around used oil containers, keeping floor drains closed when oil is being drained, and by training employees on spill prevention techniques. Oil that is contained rather than released can be recycled, thus saving money, and protecting the environment.

Spent Fluids

Farm machinery and vehicles require regular changing of fluids, including oil, coolant, and others. To minimize releases to the environment, these fluids should be drained and replaced in areas where there are no connections to storm drains or municipal sewers. Minor spills should be cleaned prior to reaching drains. Used fluid should be collected and stored in separate containers. Fluids can often be recycled.

During the process of engine maintenance, spills of fluids are likely to occur. The "dry shop" principle encourages spills to be cleaned immediately so that spilled fluid will not evaporate to air, be transported to soil, or be discharged to waterways or sewers. The following techniques help prevent and minimize the impacts of spills:

- Collect leaking or dripping fluids in designated drip pans or containers.
- Keep a designated drip pan under the vehicle while unclipping hoses, unscrewing filters, or removing other parts.

- Immediately transfer used fluids to proper containers. Never leave drip pans or other open containers unattended.

Batteries

Farm operators have three options for managing used batteries: recycling through a supplier, recycling directly through a battery reclamation facility, or direct disposal. Most suppliers now accept spent batteries at the time of new battery purchase.

While some waste batteries must be handled as hazardous waste, lead acid batteries are not considered hazardous waste as long as they are recycled. In general, recycling batteries may reduce the amount of hazardous waste stored at a farm, and thus reduce the farm's responsibilities under RCRA.

The following best management practices are recommended to prevent used batteries from impacting the environment prior to disposal:

- Place on pallets and label by battery type.
- Protect them from the weather with a tarp, roof, or other means.
- Store them on an open rack or in a watertight secondary containment unit to prevent leaks.
- Inspect batteries for cracks and leaks as they come to the farm.
- Neutralize acid spills and dispose of the resulting waste as hazardous if it sill exhibits a characteristic of a hazardous waste.
- Avoid skin contact with leaking or damaged batteries.

Machine Shop Wastes

The major hazardous wastes from metal machining are waste cutting oils, spent machine coolant, and degreasing solvents. Scrap metal can also be a component of hazardous waste produced at a machine shop. Material substitution and recycling are the two best means to reduce the volume of these wastes.

The preferred method of reducing the amount of waste cutting oils and degreasing solvents is to substitute with water-soluble cutting oils. If non-water-soluble oils must be used, recycling waste cutting oil reduces the potential environmental impact. Machine coolant can be recycled, either by an outside recycler, or through a number of in-house systems. Coolant recycling is most easily implemented when a standardized type of coolant is used throughout the shop. Reuse and recycling of solvents also is easily achieved, although it is generally done by a permitted recycler.

Most shops collect scrap metals from machining operations and sell these to metal recyclers. Metal chips that have been removed from the coolant by filtration can be included in the scrap metal collection. Wastes should be carefully segregated to facilitate reuse and recycling.

5 Best Ways for Improving Tractor Maintenance

Each and every piece of farm equipment is an important, and oftentimes expensive investment that could be built to last if cared for properly. We understand the value of how tractor servicing will preserve the value of your well-maintained and durable machinery. For many farmers, keeping a tractor or plow in good shape is just as important as a good harvest. But, in order to get the best use out of your agriculture equipment, it's crucial to take the proper steps to maintain and clean all farming equipment on a regular basis.

Depending on the type of farm you have and the type of equipment you use, these tractor maintenance tips may be repetitive. If you're looking for confirmation that you're maintaining your machinery in the best way possible or seeking out some tips for general upkeep, refer to this list of the 5 best ways to improve your tractor servicing process:

1. Every. Single. Day: Some aspects of maintenance are necessary only annually or biannually, but other steps should be practiced on a daily basis. This includes keeping your equipment in a safe place, in a garage or barn where it will be kept from harmful weather. Keep an eye on the temperature of your engine whenever your tractor is in use to make sure it's not overheating.
2. Make a Checklist: Even the most experienced farmer might occasionally miss a step, so it's important to make a list of the necessary cleaning and maintenance procedures. This checklist should be performed daily and include such procedures as checking tire inflation of tractors, counting all tools and returning them to their designated space, checking coolant and oil, and ensuring no belts have come loose.
3. Service Records: Whenever your farm machinery needs professional servicing, either for repairs or just a standard checkup, make sure you keep detailed records of what procedure was done and who did it. That way you can reference these documents in the future should the equipment need further servicing or should you plan to sell it.
4. Changing Seasons: When your tractor comes out in the spring for the first time, give it some time to adjust. The cold and

extended period of non-use can make it a little slow to start off with. While it's waking up from its winter hibernation, check the tires, steering gear, clutch and brakes, electrical system, transmission, and hydraulics.

5. Imagine It's a Ferrari: A tractor is a prized possession for any farmer so you must treat it as such. Most Ferrari owners aren't going to take their "baby" through a local car wash; they are going to take the time to carefully clean the car to ensure the best and most intricate care is being given. The same should be true for your shiny green and yellow tractor.

Preventive Maintenance

Preventive maintenance (PM) has the following meanings:

1. The care and servicing by personnel for the purpose of maintaining equipment and facilities in satisfactory operating condition by providing for systematic inspection, detection, and correction of incipient failures either before they occur or before they develop into major defects.
2. Maintenance, including tests, measurements, adjustments, and parts replacement, performed specifically to prevent faults from occurring.

The primary goal of maintenance is to avoid or mitigate the consequences of failure of equipment. This may be by preventing the failure before it actually occurs which Planned Maintenance and Condition Based Maintenance help to achieve.

It is designed to preserve and restore equipment reliability by replacing worn components before they actually fail. Preventive maintenance activities include partial or complete overhauls at specified periods, oil changes, lubrication and so on. In addition, workers can record equipment deterioration so they know to replace or repair worn parts before they cause system failure. The ideal preventive maintenance program would prevent all equipment failure before it occurs.

There is a controversy of sorts regarding the propriety of the usage "preventative."

Subgroups

Preventive maintenance can be described as maintenance of equipment or systems before fault occurs. It can be divided into two subgroups:

- planned maintenance and
- condition-based maintenance.

The main difference of subgroups is determination of maintenance time, or determination of moment when maintenance should be performed.

While preventive maintenance is generally considered to be worthwhile, there are risks such as equipment failure or human error involved when performing preventive maintenance, just as in any maintenance operation. Preventive maintenance as scheduled overhaul or scheduled replacement provides two of the three proactive failure management policies available to the maintenance engineer. Common methods of determining what Preventive (or other) failure management policies should be applied are; OEM recommendations, requirements of codes and legislation within a jurisdiction, what an "expert" thinks ought to be done, or the maintenance that's already done to similar equipment, and most important measured values and performance indications.

In a nutshell:

- Preventive maintenance is conducted to keep equipment working and/or extend the life of the equipment.
- Corrective maintenance, sometimes called "repair," is conducted to get equipment working again.

Difference Between Preventive and Predictive Maintenance

Predictive maintenance tends to include direct measurement of the item. Example, an infrared picture of a circuit board to determine hot spots while Preventive Maintenance includes the evaluation of particles in suspension in a lubricant, sound and vibration analysis of a machine.

Examples:

- An individual bought an incandescent light bulb. The manufacturing company mentioned that the life span of the bulb is 3 years. Just before the 3 years, the individual decided to replace the bulb with a new one. This is called preventive maintenance.
- On the other hand, the individual has the opportunity to observe the bulb operation daily. After two years, the bulb starts flickering. The individual predicts at that time that the bulb is going to fail very soon and decides to change it for a new one. This is called predictive maintenance.

- The individual ignores the flickering bulb and only goes out to buy another replacement light bulb when the current one fails. This is called corrective maintenance.

Corrective Maintenance

Corrective maintenance is a maintenance task performed to identify, isolate, and rectify a fault so that the failed equipment, machine, or system can be restored to an operational condition within the tolerances or limits established for in-service operations.

A French official norm defines "corrective maintenance" as maintenance which is carried out after failure detection and is aimed at restoring an asset to a condition in which it can perform its intended function (NF EN 13306 X 60-319 standard, June 2001).

Corrective maintenance can be subdivided into "immediate corrective maintenance" (in which work starts immediately after a failure) and "deferred corrective maintenance" (in which work is delayed in conformance to a given set of maintenance rules).

Bibliography

Alger, Philip L.: *Standard Handbook for Electrical Engineers*, McGraw-Hill, New York, United States, 1949.

Bedford, B.D.; Hoft, R.G.: *Principles of Inverter Circuits*, Wiley, New York, 1964.

Blundel, Stephen J.: *Magnetism A Very Short Introduction*, Oxford University Press, London, 2012.

Bose, Bimal K.: *Power Electronics and Motor Drives : Advances and Trends*, Academic Press, New York, 2006.

Chiasson, John: *Modeling and High-Performance Control of Electric Machines*, Wiley, United States, 2005.

Day, Lance; McNeil, Ian: *Biographical Dictionary of the History of Technology*, Routledge, London, 1996.

Fink, Donald G.; Beaty, H. Wayne: *Standard Handbook for Electrical Engineers*: McGraw-Hill, New York, 1999.

Froehlich, Fritz E.:*The Froehlich/Kent Encyclopedia of Telecommunications*, Marcel Dekker, New York, 1992.

Garrison, Ervan G.: *A History of Engineering and Technology*, CRC Press, New York, 1998.

Houston, Edwin J.; Kennelly, Arthur: *Recent Types of Dynamo-Electric Machinery*, P.F. Collier and Sons, New York, 1902.

Hughes, Thomas Parke: *Networks of Power: Electrification in Western society, 1880–1930,* Johns Hopkins University Press, 2011

Krishnan, R.: *Permanent Magnet Synchronous and Brushless DC Motor Drives*, CRC Press, New York, 2008.

Lander, Cyril W.: *Power electronics*, McGraw-Hill, New York City, United States, 1993.

Langsdorf, Alexander Suss: *Theory of Alternating-Current Machinery*, McGraw-Hill, New York City, United States, 1955.

Liu, Chen-Ching: *The Electrical Engineering Handbook*, CRC Press, United States, 1997.

Mattox, D. M.: *The Foundations of Vacuum Coating Technology*, Random House, Gurgaon, Haryana, 2003.

Mortensen, S. H.; Beckwith, S.: *Standard Handbook for Electrical Engineers*, McGraw-Hill, United States, 1949.

Nilsson, J W, Riedel, S. A.: *Electric Circuits*, Pearson Prentice Hall, New York City, 2007.

Nye, David E.: *Electrifying America: Social Meanings of a New Technology*, The MIT Press, United States, 1990.

Pansini, Anthony J.: *Basic of Electric Motors*, Pennwell Publishing Company, United States, 1989.

Patrick, Dale R.; Fardo, Stephen W.: *Rotating Electrical Machines and Power Systems*, Fairmont Press, Lilburn, GA 30047, United States, 1997.

Pelly, B.R.: *Thyristor Phase-Controlled Converters and Cycloconverters : Operation, Control, and Performance*, Wiley-Interscience, US., 1971.

R. S. Khandpur: *Printed circuit boards: design, fabrication, assembly and testing*, Tata McGraw-Hill, Noida, UP, India2005.

Rosenblatt, Jack; Friedman, M. Harold: *Direct and Alternating Current Machinery*, McGraw-Hill, New York, 1963.

Stölting, H. D.; Kallenbach, E.; Amrhein, W.: *Handbook of Fractional-Horsepower Drives,* Springer, New York, 2008.

Stölting, Hans-Dieter: *Handbook of Fractional-Horsepower Drives*, Springer, New York, 2008.

Vjekoslav, Damic, John, Montgomery: *Mechatronics by bond graphs: an object-oriented approach to modelling and simulation*, Springer, New York City, 2008.

Wadhwa, C.L.: *Network analysis and synthesis, New Age International,* New Delhi, India, 2008.

Weißmantel, H; Oesingmann, P.; Möckel, A.: *Handbook of Fractional-Horsepower Drives*, Springer, New York, 2008.

Index

H

I

L

M

N

O

P

R

S

T

V

W

❑❑❑